National Center for Construction Education and Research

Plumbing Level Two

Upper Saddle River, New Jersey
Columbus, Ohio

This information is general in nature and intended for training purposes only. Actual performance of activities described in this manual requires compliance with all applicable operating, service, maintenance, and safety procedures under the direction of qualified personnel. References in this manual to patented or proprietary devices do not constitute a recommendation of their use.

10 9 8
ISBN 0-13-060479-8

Preface

This volume was developed by the National Center for Construction Education and Research (NCCER) in response to the training needs of the construction and maintenance industries. It is one of many in the NCCER's standardized craft training program. The program, covering more than 30 craft areas and including all major construction skills, was developed over a period of years by industry and education specialists. Sixteen of the largest construction and maintenance firms in the United States committed financial and human resources to the teams that wrote the curricula and planned the nationally accredited training process. These materials are industry-proven and consist of competency-based textbooks and instructor's guides.

The NCCER is a non-profit educational entity affiliated with the University of Florida and supported by the following industry and craft associations:

PARTNERING ASSOCIATIONS

- American Fire Sprinkler Association
- American Society for Training and Development
- American Welding Society
- Associated Builders and Contractors, Inc.
- Associated General Contractors of America
- Association for Career and Technical Education
- Carolinas AGC, Inc.
- Carolinas Electrical Contractors Association
- Citizens Democracy Corps
- Construction Industry Institute
- Construction Users Roundtable
- Design-Build Institute of America
- Merit Contractors Association of Canada
- Metal Building Manufacturers Association
- National Association of Minority Contractors
- National Association of State Supervisors for Trade and Industrial Education
- National Association of Women in Construction
- National Insulation Association
- National Ready Mixed Concrete Association
- National Utility Contractors Association
- National Vocational Technical Honor Society
- North American Crane Bureau
- Painting and Decorating Contractors of America
- Portland Cement Association
- SkillsUSA-VICA
- Steel Erectors Association of America
- Texas Gulf Coast Chapter ABC
- U.S. Army Corps of Engineers
- University of Florida
- Women Construction Owners and Executives, USA

Some of the features of the NCCER's standardized craft training program include the following:

- A proven record of success over many years of use by industry companies.
- National standardization providing portability of learned job skills and educational credits that will be of tremendous value to trainees.
- Recognition: upon successful completion of training with an accredited sponsor, trainees receive an industry-recognized certificate and transcript from the NCCER.
- Compliance with Apprenticeship, Training, Employer, and Labor Services (ATELS) requirements (formerly BAT) for related classroom training (CFR 29:29).
- Well-illustrated, up-to-date, and practical information.

FEATURES OF THIS BOOK

Capitalizing on a well-received campaign to redesign our textbooks, NCCER is publishing select textbooks in a two-column format. *Plumbing Level Two* incorporates the design and layout of our full-color books along with special pedagogical features. The features augment the technical material to maintain the trainees' interest and foster a deeper appreciation of the trade.

Did You Know? Explains fun facts and interesting tidbits about the plumbing trade from historical to modern times.

On the Level provides helpful hints for those entering the field by presenting tricks of the trade from plumbers in a variety of disciplines.

We're excited to be able to offer you these improvements and hope they lead to a more rewarding learning experience.

As always, your feedback is welcome! Please let us know how we are doing by visiting NCCER at www.nccer.org or e-mail us at info@nccer.org.

Acknowledgments

This curriculum was revised as a result of the farsightedness and leadership of the following sponsors:

City of Phoenix, AZ
PMA/CEFGA Apprentice Program

TDIndustries
Yeager, Inc.

This curriculum would not exist were it not for the dedication and unselfish energy of those volunteers who served on the Authoring Team. A sincere thanks is extended to:

Ed Cooper
Cary Mandeville

Roger Rotundo
Dan Warnick

We would also like to thank the following reviewers for contributing their time and expertise to this endeavor:

George Benoit
Ron Braun
James Lee

Charles Owenby
Wilford Seilhamer

A final note: This book is the result of a collaborative effort involving the production, editorial, and development staff at Prentice-Hall, Inc., and the National Center for Construction Education and Research. Thanks to all of the dedicated people involved in the many stages of this project.

Contents

Intermediate Math

COURSE MAP

This course map shows all of the modules in the second level of the Plumbing curriculum. The suggested training order begins at the bottom and proceeds up. Skill levels increase as you advance on the course map. The local Training Program Sponsor may adjust the training order.

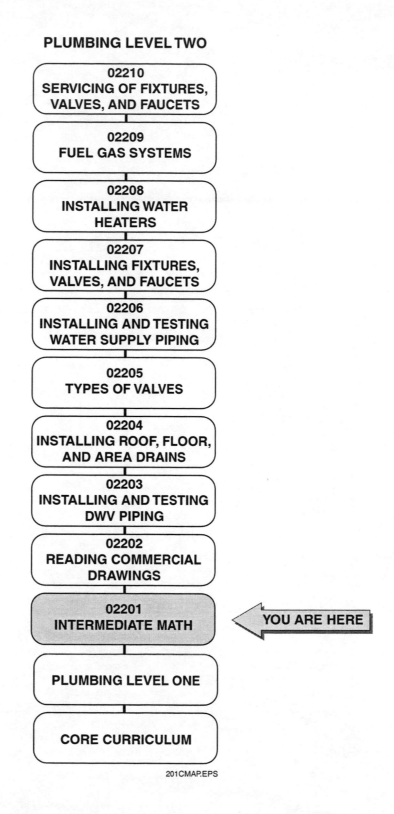

PLUMBING LEVEL TWO

02210
SERVICING OF FIXTURES, VALVES, AND FAUCETS

02209
FUEL GAS SYSTEMS

02208
INSTALLING WATER HEATERS

02207
INSTALLING FIXTURES, VALVES, AND FAUCETS

02206
INSTALLING AND TESTING WATER SUPPLY PIPING

02205
TYPES OF VALVES

02204
INSTALLING ROOF, FLOOR, AND AREA DRAINS

02203
INSTALLING AND TESTING DWV PIPING

02202
READING COMMERCIAL DRAWINGS

02201
INTERMEDIATE MATH   ⟵ **YOU ARE HERE**

PLUMBING LEVEL ONE

CORE CURRICULUM

201CMAP.EPS

Figures

Tables

Intermediate Math

Objectives

When you have completed this module, you will be able to do the following:

1. Lay out square corners using the 3-4-5 method.
2. Use a folding rule to find given angles.
3. Calculate 11¼-, 22½-, 60-, and 72-degree simple offsets.
4. Calculate 11¼-, 22½-, 60-, and 72-degree parallel offsets.
5. Calculate rolling offsets using constants for the angled fittings.
6. Use a calculator to find square root.
7. Calculate rolling offsets using a framing square.
8. Calculate 45-degree offsets around obstructions.

Prerequisites

Before you begin this module, it is recommended that you successfully complete the following modules: Core Curriculum; Plumbing Level One.

Required Trainee Materials

1. Calculator
2. Ruler or tape measure
3. Paper and sharpened pencil

1.0.0 ◆ INTRODUCTION

Building upon your previous knowledge of mathematics, this module presents **offsets** other than 45-degree offsets and introduces you to **parallel offsets** and **rolling offsets**.

2.0.0 ◆ APPLIED MATH

Math is much more than theories and formulas. When you learn to apply those theories and for-

mulas, math becomes one of the most important tools in your plumbing toolbox. In the following sections you'll learn how math helps you lay out a square corner and find angles with a folding rule.

DID YOU KNOW?

Applied and Pure Mathematics

Applied math is any math process needed to accomplish a specific task. Applied math helps you get the job done and done right. It enables you to determine the length of a run of pipe, the slope of a drain, and the number of pipe hangers you need.

Pure mathematics is the theoretical study of unproven ideas (theorems), proofs of theorems, and formulas. It is also the exploration of new fields that fit into the category of pure mathematics.

2.1.0 Laying Out a Square Corner

Math allows you to use a right triangle to make square corners. Here's how. The longest side of a right triangle squared is equal to the sums of the squares of the two shorter sides. If we call the longest leg C and the two shorter legs A and B, then we can write this relationship as a formula:

$$C^2 = A^2 + B^2$$

So how does this work in practical terms? Let's say that C = 5 feet, A = 3 feet, and B = 4 feet. Plug these numbers into the formula as follows:

$$5^2 = 3^2 + 4^2$$

Whenever the longest leg of a triangle is 5 feet and the shorter legs are 3 feet and 4 feet respectively, then the triangle is always a right triangle and contains a 90-degree angle.

Use this mathematical rule to lay out a square corner as follows:

Step 1 Using two 2 × 4s, mark one 3 feet from the end and the other 4 feet from the end as shown in *Figure 1*.

Step 2 Move the 2 × 4s in relation to each other until you can measure exactly 5 feet between the two marks.

Step 3 When you can measure exactly 5 feet between the marks, the angle must be 90 degrees. You have constructed a square corner.

This procedure works with any multiple of the 3-4-5 ratio. Thus, you could also lay out a square corner using 6 feet, 8 feet, and 10 feet.

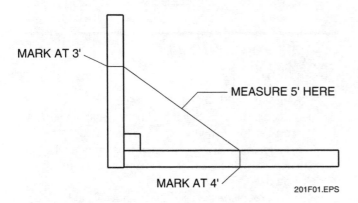

MARK AT 3'

MEASURE 5' HERE

MARK AT 4'

201F01.EPS

Figure 1 ◆ Laying out a square corner.

DID YOU KNOW?

3-4-5 Ratio

The 3-4-5 ratio is based on the Pythagorean theorem. This theorem states that in a right triangle the sum of the squares of the other two sides is equal to the hypotenuse squared ($A^2 + B^2 = C^2$). The Greek mathematician Pythagoras and his followers are credited with formulating this theorem. There is evidence that the ancient Sumerians, Egyptians, Hindus, and even the Incas and Aztecs used this same principle.

2.2.0 Finding Angles With a Folding Rule

When you are in the field, you'll sometimes need to visualize how a particular offset looks. For example, when deciding what type of angled fitting you'll need to go around an obstruction, it helps to recognize how a particular angle turns around the obstruction. The following method shows you how to use a folding rule to create angles from 11¼ degrees to 90 degrees.

To create a 90-degree angle with a folding rule, follow these steps:

Step 1 Open up the rule so that the 1-inch mark is to your left and the 24-inch mark is to your right. You will use only the first four sections of the rule.

Step 2 Moving only the first two sections of the rule, swing the upper corner of the tip of the rule up and over until the tip lines up with the lower edge of the 20¼-inch mark on the fourth section of the rule. You have created a right triangle between the first and second sections of the rule. This angle measures 90 degrees.

Step 3 Make sure the third and fourth sections are straight. Note that the angle you have created is on the *exterior* of the triangle, not inside the first and second sections.

Step 4 By holding the corner of the first section in place against the 20¼-inch mark, you can place the triangle near the obstruction to visualize the offset.

Other angles are possible, as shown in *Table 1*. Remember, only the first two sections of the rule may move. Practice forming all the angles shown in the table.

Table 1 Forming Angles

Form This Angle	Touch Tip at This Dimension
11¼°	23¹⁵⁄₁₆"
22½°	23¾"
30°	23⅜"
45°	23"
60°	22¼"
72°	21⅝"
90°	20¼"

3.0.0 ◆ Rolling Offsets

The two types of offsets you are already familiar with are the **simple offset**, shown in *Figure 2*, and the parallel offset, shown in *Figure 3*. What these offsets have in common is that the pipe remains in the same plane after it is offset. If we put an offset into a box, a simple offset would look like *Figure 4*. Notice that the run of pipe (also called the **setback**) enters the box at the bottom of one side and exits at the top of the same side.

In a rolling offset, however, something different occurs. A rolling offset results from turning a simple offset on its side (as if it were rolled over). If we put a rolling offset into a box, it would look like *Figure 5*. Notice that the rolling offset enters the box at exactly the same location as a simple offset, but exits the box from the top of a different side. The simple offset has been rolled onto its side.

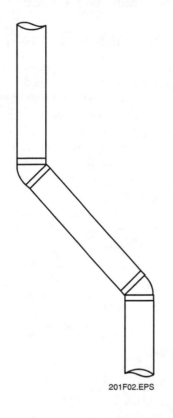

201F02.EPS

Figure 2 ◆ Simple offset.

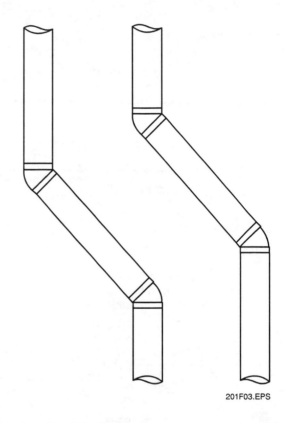

201F03.EPS

Figure 3 ◆ Parallel offset.

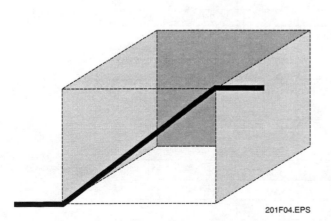

201F04.EPS

Figure 4 ◆ Simple offset in a box.

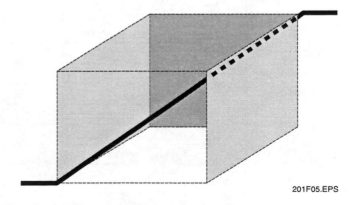

201F05.EPS

Figure 5 ◆ Rolling offset in a box.

3.1.0 Offsets—A Review

When a run of pipe changes direction, it creates an offset. Piping offsets are really right triangles. Each leg of the triangle corresponds to a component of the offset. *Figure 6* shows the individual components: offset, setback, and **travel.** Of these, travel (also called *diagonal*) is of particular importance because you must know the length of travel before you can cut the pipe to size. Once you've figured travel, subtract the takeoff for the two fittings to determine the cut size.

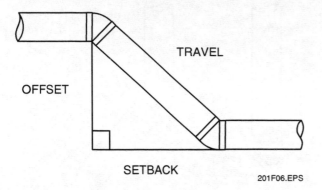

Figure 6 ◆ Offset, setback, and travel.

3.2.0 Constants

To simplify the calculations required to find travel, plumbers use constants. A constant, as its name implies, is a number that always stays the same. Each angled fitting has a constant that, when multiplied by the offset, yields the travel within the piping offset. You already know that the constant for figuring 45-degree offsets is 1.414. Multiplying the offset by 1.414 yields the travel in every 45-degree piping offset. See the Appendix for more information about constants.

3.3.0 Figuring Simple Offsets

When a run of pipe changes direction and remains in the same plane, the offset created is called a *simple offset*.

In addition to the 45-degree simple offset, you need to know how to figure 11¼-, 22½-, 60-, and 72-degree offsets. Each has its own constant. You'll figure travel in exactly the same way for each of these angles by multiplying the offset by the constant to find the travel.

Travel is really the center-to-center distance between the fittings. You need charts, like those shown in *Figures 7* and *8*, to figure the cut length.

A is Center-to-Face Measure
B is Thread-in Measure
G is Fitting-Allowance Measure

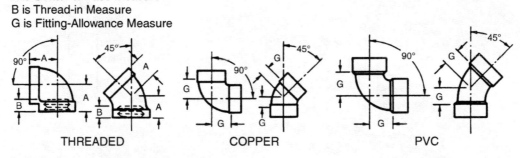

Nominal Pipe Size (inches)	Threaded			Copper		PVC	
	90° A (inches)	45° A (inches)	B (inches)	90° G (inches)	45° G (inches)	90° G (inches)	45° G (inches)
⅜	1	¾	⅜	⁵⁄₁₆	³⁄₁₆	⅜	¼
½	1⅛	¾	½	⅜	³⁄₁₆	½	¼
¾	1⅜	1	½	½	¼	⁹⁄₁₆	⁵⁄₁₆
1	1½	1⅛	½	¾	⁵⁄₁₆	¹¹⁄₁₆	⁵⁄₁₆
1¼	1¾	1⁵⁄₁₆	½	1⅛	⁷⁄₁₆	1⁹⁄₁₆	1
1½	1¹⁵⁄₁₆	1⁷⁄₁₆	½	1⁵⁄₁₆	⁹⁄₁₆	1¾	1⅛
2	2¼	1¹¹⁄₁₆	½	1⅞	¾	2⁵⁄₁₆	1½
2½	2¹¹⁄₁₆	2¹⁄₁₆	¾				
3	3¹⁄₁₆	2³⁄₁₆	1	2⅞	1⅛	3¹⁄₁₆	1¾
4	3¹³⁄₁₆	2⅝	1	3¾	1½	3⅞	2³⁄₁₆

201F07.EPS

Figure 7 ◆ Figuring cut length for 45-degree and 90-degree fittings.

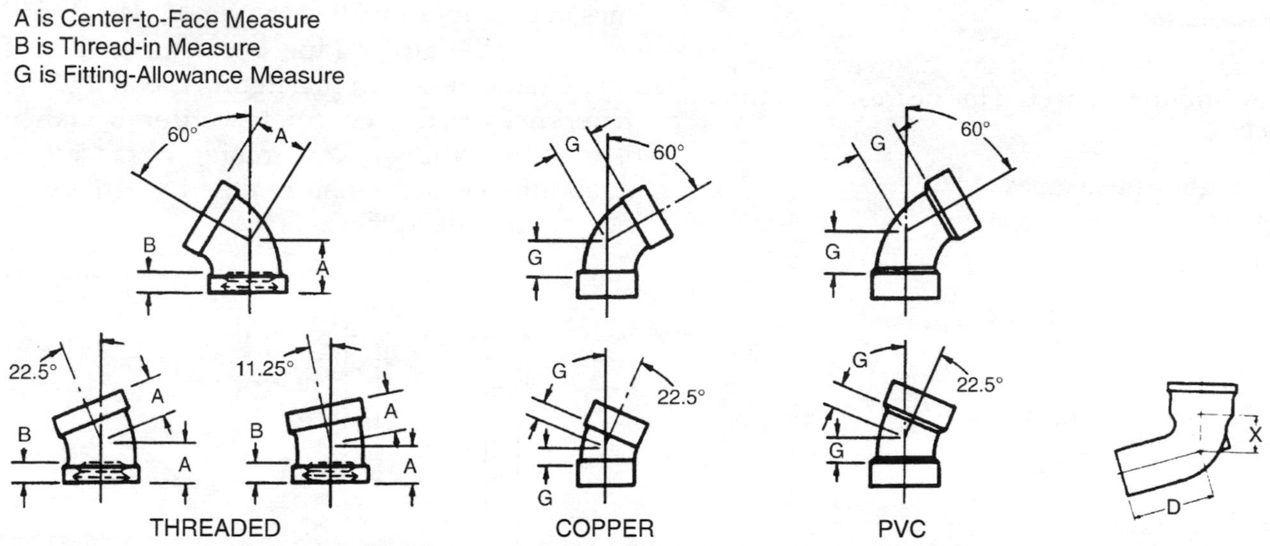

A is Center-to-Face Measure
B is Thread-in Measure
G is Fitting-Allowance Measure

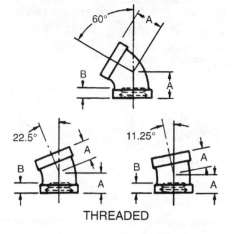

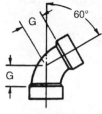

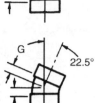
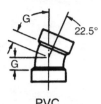

THREADED   COPPER   PVC

Nominal Pipe Size (inches)	Threaded				Copper		PVC	
	60° A (inches)	22½° A (inches)	11¼° A (inches)	B (inches)	60° G (inches)	22½° G (inches)	60° G (inches)	22½° G (inches)
1¼	1⁹⁄₁₆	1⅛	1¹⁄₁₆	½	⅝	³⁄₁₆		
1½	1¾	1¼	1¼	½	¹³⁄₁₆	¼	1	½
2	2³⁄₁₆	1½	1⅜	½	1¹⁄₁₆	⁵⁄₁₆	1⁵⁄₁₆	¹¹⁄₁₆
2½	2½	2	1⅝	¾				
3	2¹³⁄₁₆	1¹⁵⁄₁₆	1¹³⁄₁₆	1	1⁵⁄₁₆	½	1¹¹⁄₁₆	¹³⁄₁₆
4	3⅜	2⅛	2¼	1	2³⁄₁₆	¹¹⁄₁₆	2¹⁄₁₆	1

FIFTH BEND

Size	X	D	Weight
2"	2⁷⁄₁₆	5³⁄₁₆	4
3"	3¹⁄₁₆	6¹⁄₁₆	7
4"	3⁷⁄₁₆	6¹⁵⁄₁₆	11

201F08.EPS

Figure 8 ◆ Figuring cut length for 60-degree, 22½-degree, and 11¼-degree fittings.

To figure the cut length, subtract twice the center-to-face dimension (A) from the travel and add twice the thread-in (B).

3.4.0 Plumbing Constants

Table 2 shows the constants for the common angled fittings used in plumbing offsets when the offset is known. The formula for finding the travel in a piping offset when the offset is known is as follows:

Travel = Offset × Constant

Table 3 shows the constants used to find travel when the setback is known. The formula is as follows:

Travel = Setback × Constant

Table 2 Finding Travel When Offset Is Known

Angled Fitting	Constant
11¼°	5.126
22½°	2.613
45°	1.414
60°	1.1547
72°	1.05

Table 3 Finding Travel When Setback Is Known

Angled Fitting	Constant
11¼°	1.019
22½°	1.082
45°	1.414
60°	2.0
72°	3.236

Study Problems I

Find the travel for each of the following piping offsets. Round the answer to the nearest sixteenth of an inch.

1. Offset angle = 60 degrees
   Offset = 18"
   Travel = _____

2. Offset angle = 22½ degrees
   Setback = 27¾"
   Travel = _____

3. Offset angle = 11¼ degrees
   Offset = 46⅞"
   Travel = _____

4. Offset angle = 72 degrees
   Setback = 37⅝"
   Travel = _____

5. Offset angle = 45 degrees
   Offset = 14"
   Travel = _____

3.5.0 Figuring Parallel Offsets

Often, installations require two or more parallel runs of pipe, as shown in *Figure 9*. In parallel simple offsets, the distance between the center-lines of the pipes is called the *spread*. The difference in the length of Pipe 1 and the length of Pipe 2, (shown as A in this figure) is equal to the difference in the center-to-center distance between the fittings. Determining distance A is an additional calculation required when working with parallel offsets.

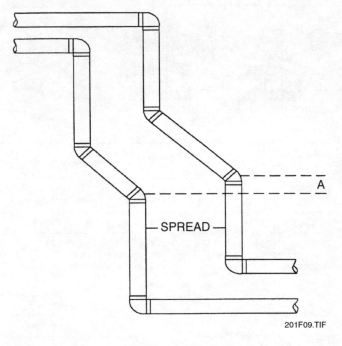

201F09.TIF

Figure 9 ◆ Parallel runs of pipe.

DID YOU KNOW?

Sir Isaac Newton (1642–1727)

Isaac Newton was one of the greatest scientists of all time. He is best known for his discovery of the law of gravity and the laws of change and motion (calculus). Much of modern science is based on the understanding and use of his laws. Although he contributed a great deal to all science, Newton is perhaps best known for his three laws of motion.

First Law of Motion—A body will rest or move with a steady speed in a straight line if other forces do not act on it. For example, if you kick a ball in outer space, it will continue to move at the same speed in a straight line if no forces such as friction or gravity act upon it. If you kick a ball on earth, gravity will pull it back to the ground and friction will eventually make the ball stop moving. Relate this law to how water is pumped from a well. The water is pumped from the well at a certain speed, but eventually gravity and friction combine to slow it down and stop it. A well-designed pump will overcome these forces.

Second Law of Motion—When a force acts on a body it produces an acceleration that is proportional to the magnitude of the force. For example, if you kick a ball, its acceleration is related to the power of your kick and the mass of the ball. The ball also moves in the direction of the force. Therefore a small, light ball travels faster than a big, heavy ball. Relate this law to how gases and fluids move through pipes. Lighter gases will move faster than heavier fluids. Fluids containing wastes will move more slowly than clear fluids.

Third Law of Motion—If A exerts a force on B, B always exerts an equal and opposite force on A. For example, if two balls traveling in opposite directions to one another collide, they will rebound back. The speed at which they collide is the sum of the two balls' speeds. Relate this law to water hammer. If air in a pipe collides with fluid in a pipe, the two materials will rebound from one another with equal force, creating a shock wave that results in water hammer.

How else does Isaac Newton's work relate to your work as a plumber? He figured out how to find the roots of numbers that you now use when calculating rolling offsets.

Remember that a parallel offset is made up of two or more simple offsets. Find the travel in each simple offset by using constants. The travel is the same length for each simple offset in a parallel offset.

3.6.0 Calculating the Difference in Pipe Legs

To find the differences in pipe legs, multiply the spread by the constant for the angled fitting. The constants for each angled fitting are shown in *Table 4*.

Difference in pipe legs = Spread × Constant

Table 4 Angled Fittings and Their Constants

Angled Fitting	Constant
11¼°	0.098
22½°	0.199
45°	0.414
60°	0.577
72°	0.727

Example:

What is the difference in pipe legs in a 22½-degree parallel offset with a spread of 12 inches?

Difference in pipe legs = Spread × Constant

= 12" × 0.199

= 2.388" or 2⅜"

Study Problems II

Find the difference in pipe legs in the following parallel offsets. Round your answer to the nearest sixteenth of an inch.

1. Angled fitting: 11¼ degrees
   Spread: 14"
   Difference: _____

2. Angled fitting: 72 degrees
   Spread: 18"
   Difference: _____

3. Angled fitting: 45 degrees
   Spread: 16¾"
   Difference: _____

4. Angled fitting: 22½ degrees
   Spread: 26⅜"
   Difference: _____

5. Angled fitting: 60 degrees
   Spread: 11⅞"
   Difference: _____

3.7.0 Laying Out a Simple Parallel Offset

To lay out a simple parallel offset, you must know two dimensions: the travel between fittings and the difference in lengths of the pipe legs. You know how to calculate each using the constants presented in this section of the module. Remember to account for the takeoff for the fittings when you lay out the offset.

3.8.0 Rolling Offsets—Terminology

You are already familiar with the terms *offset*, *travel*, and *setback* used for simple offset. Review them before studying the terminology used to describe rolling offsets. Simple offsets form a single right triangle.

The terms used to describe rolling offsets are illustrated in *Figure 10*. Rolling offsets form two right triangles. Learning how to recognize these right triangles and knowing how to work with each of them lie at the heart of figuring plumbing problems that involve rolling offsets.

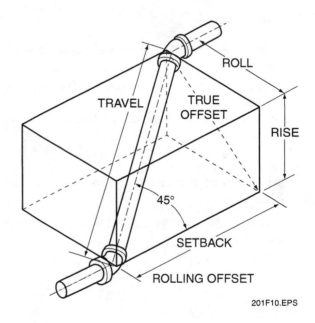

201F10.EPS

Figure 10 ◆ Rolling offset.

3.8.1 True Offset

One of the triangles formed by a rolling offset contains one leg called the *offset* and another called the **true offset** (refer to *Figure 10*). Before you can begin to solve rolling offsets, you must know the true offset.

3.8.2 Finding True Offset

A unique relationship exists among the sides of every right triangle. Because of this relationship—which can be expressed by a mathematical formula—a missing side can be found when the other two are known. The formula is expressed as follows:

$$\text{True offset} = \sqrt{\text{Rise}^2 + \text{Offset}^2}$$

If the rise is 3 feet and the offset is 4 feet, the true offset would be 5 feet:

$$\text{True offset} = \sqrt{3^2 + 4^2}$$
$$\text{True offset} = \sqrt{9 + 16}$$
$$\text{True offset} = \sqrt{25}$$
$$\text{True offset} = 5$$

The offset and **rise** form the two short legs of a right triangle. Rise is the vertical measurement; offset is the horizontal measurement. The true offset forms the long leg of the triangle.

Review Questions

Sections 2.0.0–3.8.2

1. The longest side of a right triangle squared is equal to _____.
    a. the sums of the squares of the two shorter sides
    b. the sums of the squares of the two longer sides
    c. the sum of the measurement of the two shorter sides
    d. the sum of the angles of the two shorter sides

2. The 3-4-5 ratio always creates a square corner because it forms _____.
    a. a triangle with sides that add up to 12 units
    b. a triangle with two 45-degree angles
    c. three angles inside a triangle
    d. a right triangle

3. Refer to *Table 1*. When finding angles with a folding rule, if you place the tip of the rule at 22¼ inches, it creates a _____.
    a. 22½-degree angle
    b. 45-degree angle
    c. 60-degree angle
    d. 90-degree angle

4. The center-to-center distance between fittings is called the _____.
    a. constant
    b. offset
    c. travel
    d. setback

5. The formula for finding the travel in a piping offset when the offset is known is _____.
    a. Travel = Offset × Constant
    b. Travel = Setback × Constant
    c. Travel = Setback × Offset
    d. Travel = Offset ÷ Constant

6. Refer to *Table 2*. The travel for a piping length with an offset angle of 72 degrees and an offset of 12 inches is _____. (Round the answer to the nearest sixteenth of an inch.)
    a. 12⁵⁄₁₆"
    b. 12⅝"
    c. 13¹⁄₁₆"
    d. 13⅞"

7. Refer to *Table 4*. The difference in pipe legs in a 60-degree parallel offset with a spread of 18 inches is _____. (Round the answer to the nearest sixteenth of an inch.)
    a. 25⁷⁄₁₆"
    b. 20¹³⁄₁₆"
    c. 13¹⁄₁₆"
    d. 10⅜"

8. Refer to *Table 4*. The difference in pipe legs for a 22½-degree parallel offset with a spread of 28⁷⁄₁₆ inches is _____. (Round the answer to the nearest sixteenth of an inch.)
    a. 5⅝"
    b. 11¾"
    c. 40⁵⁄₁₆"
    d. 74⅝"

9. The correct equation for figuring the true offset for an installation with a rise of 5⅝ feet and an offset of 4⅞ feet is _____.
    a. $\sqrt{5.625 + 4.875}$
    b. $\sqrt{5.625^2 + 4.875^2}$
    c. $\sqrt{5.625^2 \times 4.875^2}$
    d. $\sqrt{5.625} + \sqrt{4.875}$

10. Before you can use constants to find the travel or setback in a rolling offset, you must know the _____.
    a. true offset
    b. simple offset
    c. parallel offset
    d. angled offset

4.0.0 ◆ USING A CALCULATOR TO FIND A SQUARE ROOT

Most calculators have a square root key ($\sqrt{\ }$). To use it, first enter the number for which you want the square root. Then, press the square root key. The calculator returns the square root. If the number you entered represented feet, the decimals are decimal feet and must be converted to inches and fractions of an inch. If the number you entered was in inches, the decimals are decimal inches and must be converted to fractions of an inch. Tables are available to help you make these conversions.

After you square the offset and the rise in a rolling offset problem and add them together, you get an answer that is expressed in square units. The square root of this answer is the length of the true offset.

Study Problems III

In the following problems, find the length of the true offset by taking the square root of the given number. Express your answer in the appropriate units.

1. 16 ft. = _____

2. 24 ft. = _____

3. 132 in. = _____

4. 6.33 ft. = _____

5. 199 in. = _____

4.1.0 Finding Setback and Travel in Rolling Offsets

Once you know the true offset, you can easily find the travel and the setback by using constants.

ON THE

· LEVEL ·

The Square Root Key

This figure shows the function key used to find a square root with a calculator. Finding a square root without a calculator is a long and involved process for numbers that are more than two digits. Calculating a square root incorrectly can lead to installation errors.

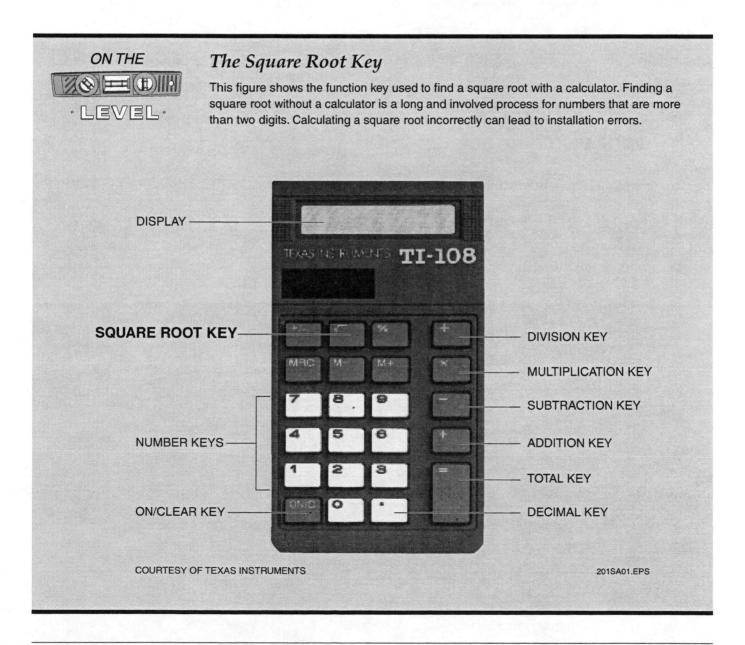

DISPLAY

TEXAS INSTRUMENTS TI-108

SQUARE ROOT KEY

DIVISION KEY

MULTIPLICATION KEY

SUBTRACTION KEY

NUMBER KEYS

ADDITION KEY

TOTAL KEY

ON/CLEAR KEY

DECIMAL KEY

COURTESY OF TEXAS INSTRUMENTS

201SA01.EPS

Remember that you used constants to help you find both simple and parallel offsets.

The constants used to find the travel and the setback in rolling offsets are shown in *Table 5*. Notice that you must know the true offset before you can use the constants to find either the travel or setback in a rolling offset. As you can see, to find travel, you multiply the true offset by the constant for the fitting angle. To find setback, multiply the true offset by the constant for the fitting angle. In each case, you use the same formula but different constants.

For example, if the fitting angle is 72: to find travel, multiply the true offset by 1.052; to find setback, multiply the true offset by 0.325.

Table 5 Constants for Rolling Offsets

Fitting Angle	72°	60°	45°	22½°	11¼°
Travel = True Offset ×	1.052	1.155	1.414	2.613	5.126
Setback = True Offset ×	0.325	0.577	1.000	2.414	5.027

5.0.0 ◆ CALCULATING ROLLING OFFSETS WITH A FRAMING SQUARE

You can also calculate rolling offsets using a framing square and tape measure or folding rule. Lay the tape measure or folding rule across the square so that it touches the rise on one side and the offset on the other. Multiply the distance between the two points by the constant for the angled fitting.

For example, if the fitting angle is 45°, to figure a rolling offset that has a 15-inch rise and an 8-inch offset, place the rule or tape as was shown in *Figure 1*. Multiply the distance measured on the rule (17 inches) by the constant for the 45-degree fitting (1.414). The answer is 24 inches, the length of the pipe center-to-center.

6.0.0 ◆ CALCULATING 45-DEGREE OFFSETS AROUND OBSTRUCTIONS

Figure 11 shows a typical situation in which a 45-degree offset must be run around an obstruction. You already know how to calculate for the simple offset; now you will learn how to determine the starting point for the offset.

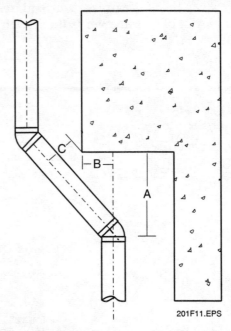

201F11.EPS

Figure 11 ◆ 45-degree offset around an obstruction.

ON THE LEVEL

Using a Framing Square

The travel measurement must be taken from inside-to-inside or outside-to-outside. Be careful not to measure from inside-to-outside of the square or from outside-to-inside.

201SA01.EPS

In the figure, you'll see three dimensions labeled A, B, and C. A is the distance from the wall to the starting point of the offset. This is what you need to figure. B is the distance from the corner to the centerline of the run of pipe. C is the distance from the corner to the centerline of the pipe.

The formula for finding the starting point of the offset is as follows:

$$A = B + (C \times 1.414)$$

For example, refer to *Figure 11*. In the 45-degree offset shown, you want the centerline of the pipe to clear the corner of the obstruction by 4 inches. You have measured the dimension marked B as 6 inches. Where should you begin the offset?

$$A = B + (C \times 1.414)$$

$$A = 6'' + (4'' \times 1.414)$$

$$A = 6'' + (5.656'')$$

$$A = 11.656'' \text{ or } 11\tfrac{5}{8}''$$

> **DID YOU KNOW?**
>
> ***Algebra***
>
> Algebra is a branch of the science of mathematics that uses numbers, letters, and other symbols and the basic operations of arithmetic (addition, subtraction, multiplication, and division) to solve problems involving unknown quantities.
>
> The constants you use when calculating simple, parallel, and rolling offsets are a basic part of algebra. Mathematics is one of the tools that you'll use every day.

Review Questions

Sections 4.0.0–6.0.0

1. A 45-degree rolling offset with an 8-inch rise and a 15-inch offset will require a pipe that measures _____ center-to-center. (Refer to Table 5.)

a. 18⅞"
b. 23"
c. 24"
d. 27⅜"

2. A 45-degree rolling offset with a 9-inch rise and a 12-inch offset will require a pipe that measures _____ center to center. (Refer to Table 5.)

a. 15"
b. 16⅞"
c. 18⅜"
d. 21¼"

3. The square root of 19⅞" = _____. (Round your answer to the nearest eighth of an inch.)

a. 4⅜"
b. 4½"
c. 4¾"
d. 4⅞"

4. A 45-degree offset must clear the corner of an obstruction by 5⅜ inches. You measure 7⅝ inches from the corner of the obstruction to the centerline of the pipe. The offset should begin _____ from the wall. (Round your answer to the nearest eighth of an inch.)

a. 9⅞"
b. 13"
c. 15¼"
d. 18⅜"

5. A 45-degree offset must clear the corner of an obstruction by 3⅝ inches. You measure 4⅜ inches from the corner of the obstruction to the centerline of the pipe. The offset should begin _____ from the wall. (Round your answer to the nearest eighth of an inch.)

a. 5⅛"
b. 9½"
c. 9⅛"
d. 11⅜"

Study Problems IV

Solve the following rolling offsets. Round the decimals to the nearest sixteenth of an inch.

1. Fitting angle: 72 degrees
Rise: 27¾"
Offset: 36"
True offset: __________
Travel: __________
Setback: __________

2. Fitting angle: 11¼ degrees
Rise: 31¾"
Offset: 40"
True offset: __________
Travel: __________
Setback: __________

3. Fitting angle: 22½ degrees
Rise: 30¾"
Offset: 39"
True offset: __________
Travel: __________
Setback: __________

4. Fitting angle: 60 degrees

   Rise: 28¾"

   Offset: 37"

   True offset: _____

   Travel: _____

   Setback: _____

5. Fitting angle: 45 degrees

   Rise: 29¾"

   Offset: 38"

   True offset: _____

   Travel: _____

   Setback: _____

Study Problems V

For each of the following 45-degree simple offsets around an obstruction, find where the offset should begin. Refer to *Figure 11* as a guide.

1. A = _____

   B = 8"

   C = 3½"

2. A = _____

   B = 6"

   C = 5"

3. A = _____

   B = 14"

   C = 4"

4. A = _____

   B = 11½"

   C = 4¾"

5. A = _____

   B = 16⅝"

   C = 6⅞"

Summary

Math is one of the most important tools in your plumber's toolbox. More than just theories and formulas, math helps you solve practical problems in the field. Tricks of the trade, such as using triangles to square corners or using a folding rule to find angles, help make you a better, more efficient plumber. Among the math-related tools you'll use are the folding rule, the framing square, and the calculator.

When a run of pipe changes direction, it creates an offset; you learned how to calculate simple, parallel, and rolling offsets, as well as how to calculate 45-degree offsets around obstructions.

You'll run into math at every stage of plumbing, whether you use it to figure lengths of pipe, to calculate offsets, or to square a corner. Studying math and applying it every day is a career-long process.

Trade Terms Introduced in This Module

Offset: A condition created when a run of pipe changes direction.

Parallel offset: A condition in which the run of pipe remains in the same plane after it is offset.

Rise: The vertical measurement as opposed to the horizontal measurement of a run pipe.

Rolling offset: A condition that results from turning a simple offset on its side (as if it were rolled over).

Setback: The distance between where a run of pipe starts to change direction and where it returns to its original direction.

Simple offset: A condition in which the run of pipe changes direction.

Travel: The longest leg of a right triangle. It is also the center-to-center distance between fittings.

True offset: The longest of the legs of a triangle formed by a rolling offset.

Constants

This appendix is included for those who wish to pursue the subject of constants further. It explains how the constants for the various angled fittings are determined.

The Constants for Simple Offsets

The constants you were given in this module to help you figure the travel for simple piping offsets have their source in trigonometry.

Piping offsets form right triangles. In the right triangle, the travel is the longest leg of the right triangle, also known as the hypotenuse. The offset is the leg of the right triangle that is opposite the angle formed by the fitting.

In the right triangle, both the offset and the angle of the fitting are known. What's needed is the travel—the hypotenuse of the triangle. In trigonometry, you find the cosecant of an angle by dividing the hypotenuse by the side opposite the angle. This means that the side opposite the angle (the offset) times the cosecant of the angle yields the value of the hypotenuse (travel).

You can find the cosecant of the various angles from trig tables or with a calculator that has trigonometric functions. If you have a calculator with trig functions, you can find the cosecant of an angle by entering the angle and then pressing the key marked *sin*. The number displayed is the sine of the angle. To find the cosecant, press the key labeled *1/x*. This gives you the reciprocal of the value. Because the reciprocal of the sine is the cosecant, the cosecant for the angle is displayed.

You can derive the constant for finding the travel when you know the setback from the cotangent function in trigonometry. This function is the reciprocal of the tangent function. You can find this value by pressing the *1/x* key after you have found the tangent of a given angle.

The Constants for Parallel Offsets

Parallel offsets have parallel angles that are one-half the offset angle. For example, in a 45-degree parallel offset, the 45-degree fittings that make up each of the simple offsets are 22½ degrees in relation to each other.

You can derive the constants for parallel offsets from the tangent function in trigonometry. They represent the tangent of one-half the fitting angle. You can find the tangent by dividing the side opposite the angle by the side adjacent to the angle.

To find the constant for parallel offsets by using a calculator with trig functions, first enter the angle of the offset. Then divide the angle by two. Finally, press the *tan* key.

Answers to Review Questions

Sections 2.0.0–3.8.2
1. a
2. d
3. c
4. c
5. a
6. b
7. d
8. a
9. b
10. a

Sections 4.0.0–6.0.0
1. c
2. d
3. b
4. c
5. b

Answers to Study Problems

Study Problems I
1. 20.7846" or 20¹³⁄₁₆"
2. 30"–30.0255"
3. 240⁵⁄₁₆" or 240¼" rounded up
4. 120.7433" or 120¾" rounded down
5. 19¹³⁄₁₆" or 19¾" rounded down

Study Problems II
1. 1⅜"
2. 13¹⁄₁₆"
3. 6¹⁵⁄₁₆" or 7" rounded up
4. 5¼"
5. 6⅞"

Study Problems III
1. 4 ft.
2. 4.898979 ft.
3. 11.48913 in.
4. 2.515949 ft.
5. 14.10674 in.

Study Problems IV
1. True offset: 45.45395 or 45⁷⁄₁₆"
   Travel: 47.81757 or 47¹³⁄₁₆"
   Setback: 14.77253 or 14¾"
2. True offset: 51.0692 or 51¹⁄₁₆"
   Travel: 261.7807 or 261¾"
   Setback: 256.7249 or 256¾"
3. True offset: 49.6645 or 49¹¹⁄₁₆"
   Travel: 129.7733 or 129¾"
   Setback: 119.8901 or 119⅞"
4. True offset: 46.85683 or 46⅞"
   Travel: 54.11964 or 54⅛"
   Setback: 27.03639 or 27¹⁄₁₆"
5. True offset: 48.26 or 48¼"
   Travel: 68.23964 or 68¼"
   Setback: 48.26 or 48¼"

Study Problems V
1. 12¹⁵⁄₁₆"
2. 13¹⁄₁₆"
3. 19⅝"
4. 18³⁄₁₆"
5. 26⅜"

NCCER CRAFT TRAINING USER UPDATES

The NCCER makes every effort to keep these textbooks up-to-date and free of technical errors. We appreciate your help in this process. If you have an idea for improving this textbook, or if you find an error, a typographical mistake, or an inaccuracy in the NCCER's Craft Training textbooks, please write us, using this form or a photocopy. Be sure to include the exact module number, page number, a detailed description, and the correction, if applicable. Your input will be brought to the attention of the Technical Review Committee. Thank you for your assistance.

Instructors – If you found that additional materials were necessary in order to teach this module effectively, please let us know so that we may include them in the Equipment and Materials list in the Instructor's Guide.

Write: Curriculum Revision and Development Department
National Center for Construction Education and Research
P.O. Box 141104, Gainesville, FL 32614-1104

Fax: 352-334-0932

E-mail: curriculum@nccer.org

Craft _____ Module Name _____

Copyright Date _____ Module Number _____ Page Number(s) _____

Description _____

_____

_____

_____

(Optional) Correction _____

_____

_____

(Optional) Your Name and Address _____

_____

_____

Reading Commercial Drawings

COURSE MAP

This course map shows all of the modules in the second level of the Plumbing curriculum. The suggested training order begins at the bottom and proceeds up. Skill levels increase as you advance on the course map. The local Training Program Sponsor may adjust the training order.

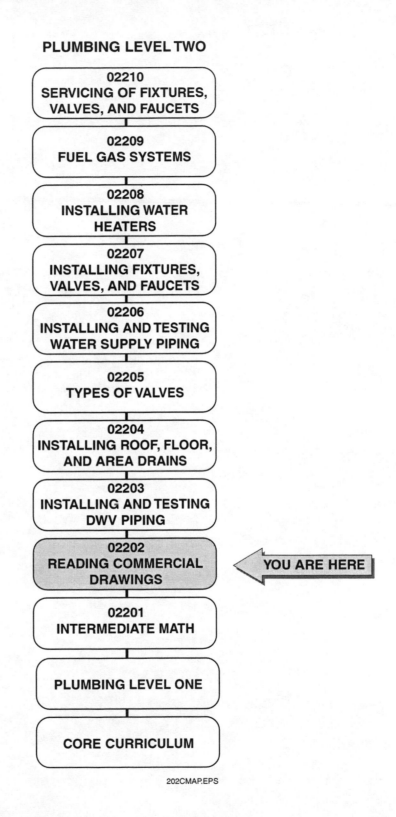

PLUMBING LEVEL TWO

02210
SERVICING OF FIXTURES, VALVES, AND FAUCETS

02209
FUEL GAS SYSTEMS

02208
INSTALLING WATER HEATERS

02207
INSTALLING FIXTURES, VALVES, AND FAUCETS

02206
INSTALLING AND TESTING WATER SUPPLY PIPING

02205
TYPES OF VALVES

02204
INSTALLING ROOF, FLOOR, AND AREA DRAINS

02203
INSTALLING AND TESTING DWV PIPING

02202
READING COMMERCIAL DRAWINGS ⟵ YOU ARE HERE

02201
INTERMEDIATE MATH

PLUMBING LEVEL ONE

CORE CURRICULUM

202CMAP.EPS

MODULE 02202 CONTENTS

Figures

Tables

Reading Commercial Drawings

Objectives

When you complete this module, you will be able to do the following:

1. Interpret information from given site plans.
2. Verify dimensions shown on drawings and generate a Request for Information (RFI) when you find discrepancies.
3. Locate plumbing entry points, walls, and chases.
4. Create an isometric drawing.
5. Do a material takeoff for drainage, waste, and vent (DWV) and water supply systems from information shown on drawings.
6. Use cut sheets and floor plans to lay out fixture rough-ins.

PREREQUISITES

Before you begin this module, it is recommended that you successfully complete the following modules: Core Curriculum; Plumbing Level One; Plumbing Level Two, Module 02201.

1.0.0 ◆ INTRODUCTION

Commercial projects are more complicated than residential projects. The buildings may be multistory. Different construction trades are on the site at different times. Often the work of one trade can't begin until another trade successfully completes its work. Commercial **plumbing drawings** likewise are more complex than residential plumbing drawings. This module introduces you to the information you'll find in a typical set of commercial drawings, with an emphasis on how it relates to your work as a plumber.

2.0.0 ◆ OVERVIEW OF COMMERCIAL DRAWINGS

A typical set of commercial blueprints includes a variety of drawings. Each type of drawing is referred to by name, and the pages usually are coded to identify the drawing with a particular category. Common categories of drawings are as follows:

- Civil–C
- Architectural–A
- Structural–S
- Mechanical–M
- Plumbing–P
- Electrical–E

Depending on the size of the project, other types of drawings may be included. For example, commercial office buildings often include **landscape drawings**. These drawings are designated with the letter L.

Generally, there is a standard way to read and use blueprints. However, some companies may have developed some ways—unique to them—to denote elements of a blueprint. Don't expect all blueprints to follow the same procedure outlined in this module. Learn your company's procedures as well.

2.1.0 Civil Drawings—The Site Plans

The **civil drawings**, also called the **site plans**, describe the overall shape of the building site. In addition, these drawings show a building's location and other structures planned for the site. Pages on these drawings are numbered C-1, C-2, and so on.

Most site plans are drawn to a relatively small scale—for example, 1" = 20'-0". A small scale generally is used so that the entire site can be shown on one page. However, the small scale limits the

ON THE

Studying Plans

When you look at a **floor plan**, first locate any rooms that require plumbing: kitchen, bathrooms, and utility rooms. Review these areas to determine the piping and fixtures needed for each. Then study the rest of the floor plan carefully for other areas that might require water supply or DWV piping.

amount of information that can be shown. Therefore, other drawings that provide sufficient detail are included in the set.

Plumbers use civil drawings to locate existing utilities.

2.2.0 Architectural Drawings

Architectural drawings show the shape, size, and appearance of the structure. They include the plan views, **elevations**, sections, and details necessary to convey information about the structure. Pages on these drawings are numbered A-1, A-2, and so on.

Plumbers use the floor plans from architectural drawings to rough-in the number, location, and spacing of plumbing fixtures.

2.3.0 Structural Drawings

Structural drawings show how the framework of the building is to be constructed. This may include structural steel, cast-in-place concrete, precast concrete, or other material. Pages on these drawings are numbered S-1, S-2, and so on.

The location of structural piping supports affects where you can install pipes, so you'll need to refer to these drawings. You may have to use different kinds of structural piping supports to secure pipe hangers and connectors.

2.4.0 Mechanical Drawings

Mechanical drawings show the location and size of the components of the heating, ventilating, and air-conditioning (HVAC) system. They also may show elevators, conveyors, and escalators. Pages on these drawings are numbered M-1, M-2, and so on.

Check the mechanical drawings to determine what limitations mechanical equipment will impose on the plumbing system. You may be required to install the piping that connects to some of the mechanical equipment.

2.5.0 Plumbing Drawings

Plumbing drawings show the location and size of the drain, waste, and vent (DWV) and water supply piping. These drawings include a special floor plan. These floor plans omit the detailed information needed by other trades and provide additional details plumbers need. Pages on these drawings are numbered P-1, P-2, and so on.

Frequently, **riser diagrams** are included on plumbing drawings to show the installation of the vertical piping system (see *Figure 1*).

2.6.0 Electrical Drawings

Electrical drawings show the details of the electrical system. They typically show the location of fixtures, controls, receptacles, circuit breakers, and wiring for larger electrical devices. Pages on these drawings are numbered E-1, E-2, and so on.

 DID YOU KNOW?

A Museum for Architecture

The Athenaeum of Philadelphia is a museum of American architecture. The museum, which has collected the work of about 1,000 American architects, has 150,000 drawings, 50,000 photographs, and many architectural documents.

ON THE

Plan Size

The drawings for a small residence may be combined. The electrical plan, for example, may be included on the floor plan. So there may be no more than three or four plan sheets. In contrast, the plans for a large commercial building may contain more than 100 pages. Each page shows a different aspect of the plan—the floor plan, the electrical plan, the plumbing plan, the HVAC plan, and so on.

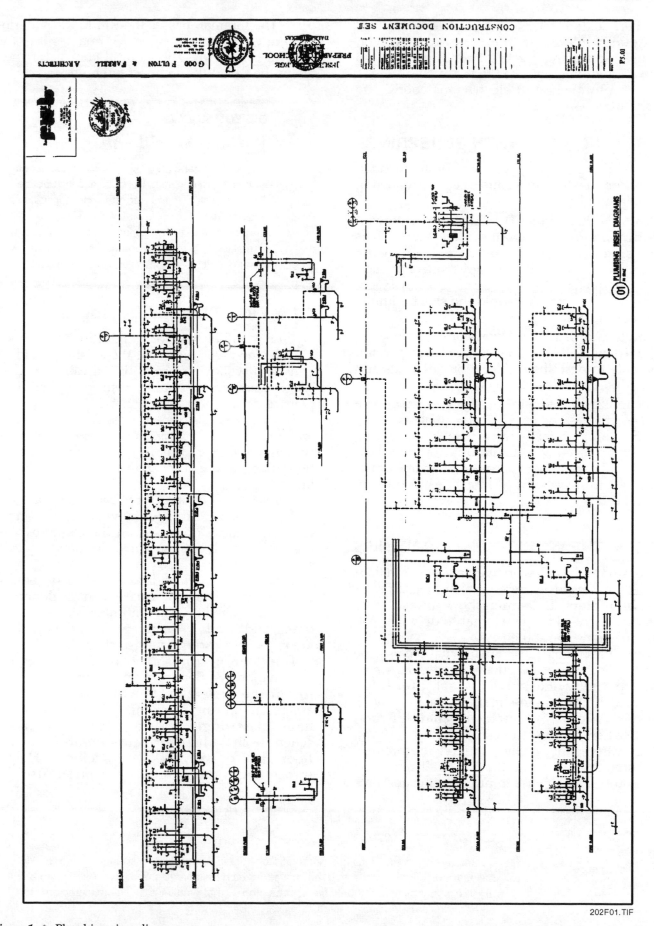

Figure 1 ◆ Plumbing riser diagram.

Plumbers seldom refer to the electrical drawings for details related to their work. However, because piping and electrical conduit often run through buildings in areas that are close to one another, plumbers and electricians coordinate their work.

3.0.0 ◆ WORKING WITH BLUEPRINTS

When working with any type of drawing found in a set of commercial blueprints, keep the following points in mind:

- Read the **title block**. The title block gives you critical information about the drawing such as the scale, the last revision date, the drawing number, and the name of the architect or engineer. If you have to remove a sheet from a set of drawings, be sure to re-fold the sheet with the title block facing up.
- Find the **north arrow**. Always orient yourself to the structure. If you know where north is, you'll be able to describe the locations of walls and other parts of the building accurately.
- Recognize that blueprints work together as a set. Architects and engineers draw plans, elevations, and sections because it takes more than one type of view to communicate the whole project. Learn how to use more than one drawing, when necessary, to find the information you need.

4.0.0 ◆ THE WORKSHEET DRAWINGS

In this module, you'll work with a site plan, a floor plan, plumbing drawings, and isometric drawings. In addition to learning how to read these plans, you'll need to know what to do when you have questions about them.

Sometimes plans may contain what seems to be an error. Sometimes questions about details may come up. Craftworkers use a **Request for Information (RFI)** for these situations (see *Figure 2*). Workers follow a hierarchy, or chain-of-command, when issuing RFIs. For example, if you notice a discrepancy on the plans, tell the foreman. The foreman will write the RFI, using specific details and making sure to put the time and date on it. The foreman passes the RFI to the superintendent or project manager, who passes it to the general contractor. The general contractor then relays the RFI to the architect or engineer.

DID YOU KNOW?

William Strickland

William Strickland (1787–1854) was an American architect and engineer. His buildings include the Second Bank of the United States in Philadelphia and the State Capitol in Nashville, Tennessee. His work combined the elegance of ancient Greece with efficient floor plans and modern materials like cast iron, steel, reinforced concrete, and glass.

4.1.0 Floor Plan and Schedules

Figure 3 is a floor plan for an addition to a school. The shaded areas represent the existing building and front walkway. The addition calls for more classrooms and two public restrooms. The floor plan describes the size and shape of the floor and includes necessary dimensions. Floor plans generally are drawn to a scale of ⅛" = 1'-0", although sometimes ¼" = 1'-0" is used. Plumbers refer to floor plans for the number and location of plumbing fixtures.

Figure 3 also contains three **schedules:** a floor-and-finish schedule, a window schedule, and a door schedule. Depending on the complexity of the project, each of these schedules may require a separate page or pages.

Designers use a standard marking system, or convention, to show information on the floor plan and in the schedule. They number all the rooms, doors, and windows on the plan. These numbers are keyed to a corresponding schedule, or list.

The floor-and-finish schedule provides the information that floor-and-finish contractors need to put the project out for bids. The designer assigns room numbers (shown in boxes) to the halls, classrooms, and restrooms. The finish schedule lists the treatments required for the floors, baseboards, walls, and ceilings. Height information and a remarks section also are included.

ON THE **LEVEL**

Changing Plans

Sometimes plumbing plans are incorrect. Sometimes they have to be changed after building has started. In these cases, you may have to sketch the changes. You don't need to be a fine artist, but your sketches should be clear enough so that architects, building inspectors, and other trades can understand them.

**XYZ, Inc
General Contractors
123 Main Street
Bigtown, USA 10001
(111) 444-5555**

PROJECT:

TO:

R.F.I.

Request for Information

**XYZ Project #_____
Date: _____
R.F.I. # _____**

RE:

Specification Reference: _____

Drawing Reference: _____

SUBJECT:

REQUIRED:

Date Information Is Required: _____

XYZ, Inc By: _____

REPLY:

**Distribution: Superintendent
              Field File**

By: _____

Date: _____

202F02.TIF

Figure 2 ◆ Sample Request for Information (RFI).

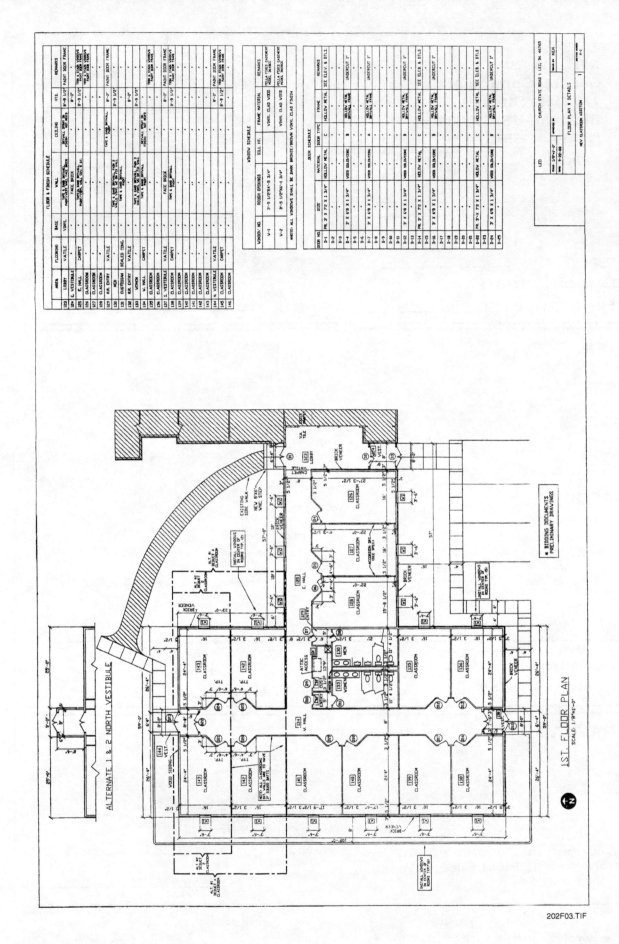

Figure 3 ◆ Floor plan and schedules.

The Detail Plan

Whether you are plumbing a commercial building or a residence, one detail drawing you'll find useful is the kitchen cabinet plan. You can determine the location and type of sink from this drawing. In addition, if a window is planned over the kitchen sink, for example, you'll need to adjust the location of the vent stack. While kitchens and baths in most modern commercial buildings have no exterior walls or windows, you may still encounter this situation.

Doors are numbered sequentially. The designer draws a circle in front of each door location and writes the appropriate number there. You can compare the circled number in the drawing to its corresponding number in the schedule to get details about the door—height and width, materials, frame—and any additional notes the designer makes. For example, door D-16 at the entrance to classroom 138 is a wood, solid core door with a hollow metal drywall frame. The size of the door is always shown as width by height by thickness. See the detail of the door schedule in *Table 1*.

For windows, designers create a system to show the size of each window and where to install it. In this example, the design calls for two sizes of window, W-1 and W-2. The designer writes those numbers in boxes along the outer walls of the addition at each window location. Window suppliers and installers use this schedule to provide the right number of windows and to match the correct size to the marked opening. The window schedule contains information about the rough opening, the frame material, the model number, and the manufacturer's name. No information appears in the column titled *sill height* because these drawings are

marked for bidding only. See *Table 2* for a detail of this schedule. Note that the restrooms have no outside walls, so the designer has not included any windows. Generally, restrooms in commercial buildings do not include windows. However, other rooms requiring plumbing, such as kitchens, may have windows. Windows in rooms that require plumbing can affect the position of the vent stack.

Figure 3 also contains two **alternates**. In this example, the owner wants to see how much the project will cost if fewer classrooms are included. So Alternate 1 requires a bid on the project with four fewer classrooms, and Alternate 2 requires a bid with two fewer classrooms. This way, the owner receives three bids as follows:

• The full project
• The full project with two fewer classrooms
• The full project with four fewer classrooms

By requesting alternate bids, the owner can plan more effectively, stay on budget, and get the most classrooms for the budgeted amount.

Alternates can affect the plumbing for a project; therefore, you must be aware of them.

Table 1 Door Schedule

Door No.	Size	Material	Door Type	Frame	Remarks
D-14	PR 3' × 7'-2" × 1¾"	Hollow Metal	C	Hollow Metal	See Elev & Dtls.
D-15	"	"	"	"	"
D-16	3' × 6'-8" × 1¾"	Wood/Solid Core	B	Hollow Metal Drywall Frame	Undercut 1"

Table 2 Window Schedule

Window No.	Rough Openings	Sill Ht.	Frame Material	Remarks
W-1	3'-5½" × 4'-5¾"		Vinyl-clad wood	Pella fixed casement Model 3648CC
W-2	3'-5½" × 6'-1¾"		Vinyl-clad wood	Pella fixed casement Model 3668CC
	Windows shall be dark bronze/brown vinyl clad finish.			

4.1.1 Entry Points, Walls, and Chases

To locate the plumbing entry points, study the floor plan. Trace a route from the building sewer to the main stack to the restroom. Architects generally show exterior walls as two lines, drawn side by side. These run around the perimeter of the building.

Interior walls are drawn as a single, solid line. Rooms on a floor plan are easy to identify. They usually include a doorway and in many cases are labeled with the name or number of the room.

In commercial restrooms, the **chase** is a hollow, enclosed area built between two back-to-back restrooms. All the plumbing connections fit into this area.

4.2.0 The Sanitary Plumbing Plan

The sanitary plumbing plan has some information in common with the floor plan of the building. The overall dimensions are the same as those on the floor plan, but the sanitary plumbing plan does not include detailed interior dimensions.

This figure also contains an **isometric drawing** of the sanitary and vent system. Notice how the piping sizes are represented. The drawing also contains a schedule of plumbing systems. This schedule establishes the governing plumbing codes, specifies piping materials, and calls out insulation requirements.

You learned how to make isometric drawings in Plumbing Level One. But you may want to refresh your memory. Study the steps for making an isometric sketch shown in *Figure 4*.

Below the schedule of plumbing systems in *Figure 5*, you'll find a section on plumbing system specifications. This section contains information about installing and testing the systems. To lay out and install the fixtures shown on the drawing, you must get fixture rough-in sheets from the purchasing agent or project manager.

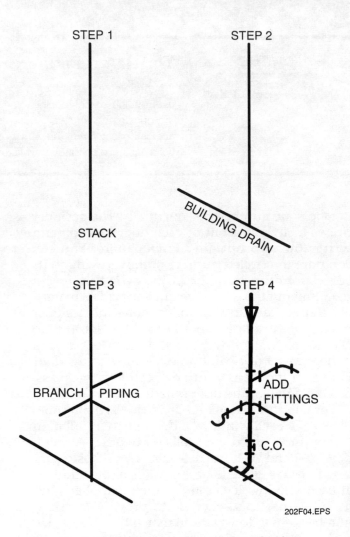

Figure 4 ◆ Steps for making an isometric sketch.

4.3.0 Plumbing Plan for the Water Supply System

Figure 6 shows the plumbing requirements for the water supply system. Compare this drawing with the drawing in *Figure 5* and the site plan for *Practice Problem 1* to get a better idea of the plumbing requirements for this building.

DID YOU KNOW?
Isometric Drawings

Isometric means equal measurement. A designer uses the true dimension of an object to construct the drawing. An isometric drawing shows a three-dimensional view of where the pipes should be installed. The piping may be drawn to scale or to dimension, or both. In an isometric drawing, vertical pipes are drawn vertically on the sketch, and horizontal pipes are drawn at an angle to the vertical lines.

DID YOU KNOW?
The Empire State Building

The Empire State Building in New York City was, at 102 stories, the tallest building in the world until 1972 (when the World Trade Center took that title). Workers took only 14 months to build the Empire State Building. It has 70 miles of water pipe that provide water to tanks at various levels. The highest tank is on the 101st floor. There are two public restrooms on each floor and a number of private bathrooms as well. Interestingly, there are no water fountains.

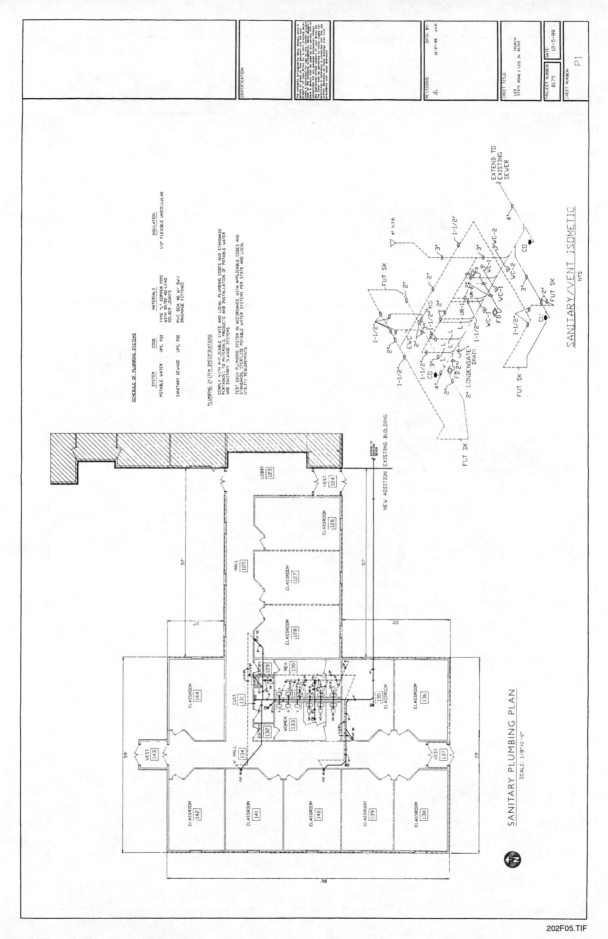

Figure 5 ◆ The sanitary plumbing plan.

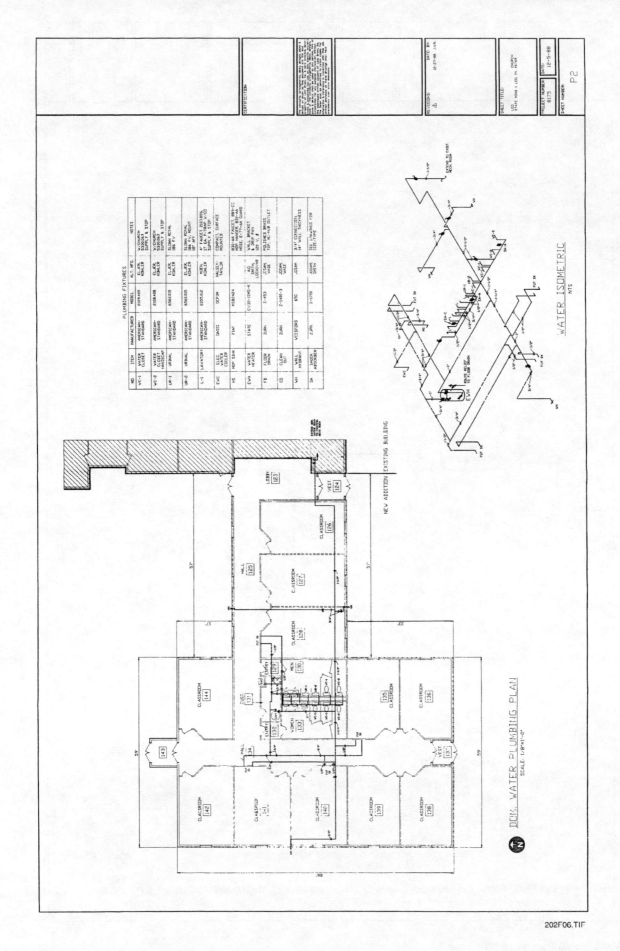

Figure 6 ◆ Plumbing plan for the water supply system.

Note that this plan contains an isometric drawing for the water supply. This drawing shows the general route of the water supply piping, the piping sizes, and the general locations of the urinals, water closets, and lavatories.

The drawing also includes a schedule of plumbing fixtures. The first column contains the letter/number key that corresponds to the fixture's location in the isometric drawing. Also included are the type of fixture, the manufacturer's name, the model number, an alternate manufacturer, and notes. Generally, the owner will specify the name of the manufacturer and the model wanted. The plumbing fixtures schedule may include a list of alternates of equal value. This gives the contractor or plumber more flexibility in getting fixtures to complete the job. See *Table 3* for a detail of the plumbing fixtures schedule. Important details about fixtures are shown on a **cut sheet**, which is a drawing of a specific fixture that includes rough-in dimensions, the manufacturer's specifications, and other information (see *Figure 7*).

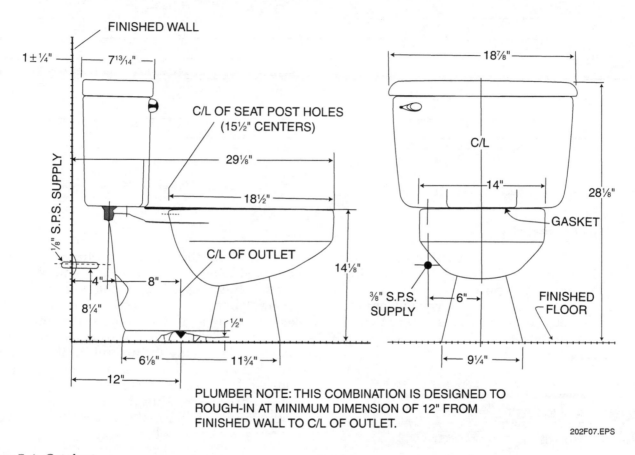

Figure 7 ◆ Cut sheet.

Table 3 Plumbing Fixtures Schedule

No.	Item	Manufacturer	Model	Alt. MFG.	Notes
WC-1	Water closet	American Standard	2109.405	Eljer, Kohler	W/Church 5330.063 Supply & Stop
WC-2	Water closet Handicap	American Standard	2108.408	Eljer, Kohler	W/Church 5330.063 Supply & Stop
UR-1	Urinal	American Standard	6560.015	Eljer, Kohler	Sloan Royal 186 FV

Project:				Estimate No.		
Location:				Sheet No.		
Architect/Engineer;				Date:		
Summary by:			Prices by:		Checked by:	
Quantity	**Description**	**Unit Price Material Cost**	**Total Estimated**	**Unit Price Labor Cost**	**Total Estimated**	**Total**

202F08.EPS

Figure 8 ◆ Material takeoff.

4.4.0 The Material Takeoff

A material takeoff is like a shopping list for the job. Use the isometric drawings and plumbing drawings to determine how many of each item (fixtures, piping, and other items) you will need to complete the job. You'll include a description of each item and an estimate of material and labor costs (see *Figure 8*).

DID YOU KNOW?

R. Buckminster Fuller

R. Buckminster Fuller (1895–1983) was an architect and engineer. He is best known for the geodesic dome. This is the lightest, strongest, and most cost-effective structure ever devised. It can enclose a large area without internal supports. In 1957, a geodesic dome auditorium built in Honolulu, Hawaii, held its first concert about 22 hours after delivery of the building parts. In the 70s and 80s, the geodesic dome became popular in some residential markets.

Summary

To understand the scope of a plumbing project, you must know how to read and interpret commercial drawings. These include civil, architectural, structural, mechanical, plumbing, and electrical drawings. Each of these drawings gives you different, yet related, information. These drawings, especially the architectural, structural, and plumbing drawings, affect how you will do your work. To become an accomplished blueprint reader, practice reading these plans. Blueprint reading is a tool, like mathematics, that you need in your plumber's toolbox.

REVIEW QUESTIONS

1. A civil drawing shows the _____.
   a. shape of the building site, the building location, and other structures planned
   b. shape of the building site, the floor plan, and a schedule of elevations
   c. topography of the site, the HVAC plan, and the plumbing and piping plan
   d. property lines, a list of the finish schedules, and the governing codes

2. Most site plans are drawn to a scale of _____.
   a. 1" = 1'-0"
   b. 1/8" = 1'-0"
   c. 1" = 20'-0"
   d. 1" = 10'-0"

3. Plumbers must be familiar with structural drawings because they show _____.
   a. where the structural members might affect the piping
   b. where they are allowed to modify the structural members
   c. what types of hangers and piping support to use
   d. what types of structural members must not be modified

4. You will find information about the HVAC system on the _____.
   a. civil drawings
   b. structural drawings
   c. plumbing drawings
   d. mechanical drawings

5. You should know how to read mechanical drawings because _____.
   a. you will have to locate and size the HVAC units
   b. they are included in your set of plumbing plan drawings
   c. you may have to install the piping that connects to the HVAC units
   d. you may have to review them for errors

6. The title block on blueprints shows the _____.
   a. north arrow
   b. last revision date
   c. number of elevations
   d. current date

7. If you find an error on a drawing, you should _____.
   a. write out the error on an RFI clearly and carefully and date it
   b. write up an RFI, give two copies to the architect, and keep one for yourself
   c. explain the problem to your foreman, who writes the RFI
   d. send an RFI by registered mail to the general contractor

8. Alternate plans are important to plumbers because they _____.
   a. can affect all aspects of plumbing for a job
   b. might delay the start of plumbing work
   c. must be approved by the plumbing inspector
   d. can increase the number of plumbing tests required

9. To lay out and install the plumbing fixtures shown on a drawing, you must _____.
   a. study the isometric drawing for proper spacing of fixtures
   b. get the manufacturer's catalog of fixtures and a comparative price list
   c. do a material takeoff for the fixtures and have it approved by the foreman
   d. get the fixture rough-in sheets from the purchasing agent or project manager

10. A cut sheet is _____.
    a. a detail plan cut from the original blueprints
    b. a plan showing where plumbers may cut into structural members
    c. a detachable page from a material takeoff
    d. a drawing of a specific fixture with rough-in dimensions

PRACTICE PROBLEMS

Use the following practice problems to answer questions about the site plan; to generate an RFI; to locate information about plumbing fixtures; to locate the DWV system on the isometric drawing; to do material takeoffs; and to complete other exercises that may be assigned. Use the figures that accompany the problems and *Appendix A* and *B*.

Practice Problem I: The Site Plan

The illustration for *Practice Problem I* shows a site plan for an addition to a school. Use this plan to answer the following questions.

1. What is the scale?

   _____

2. Where would you find the client's name?

   _____

3. How can you tell which part of the drawing shows the existing building and which shows the proposed addition?

   _____

4. Locate the north arrow. Which wall of the existing building—north, south, east, or west— does the new classroom addition tie into?

   _____

5. What type of line designates the property line? Where did you find this information?

   _____

   _____

   _____

   _____

   _____

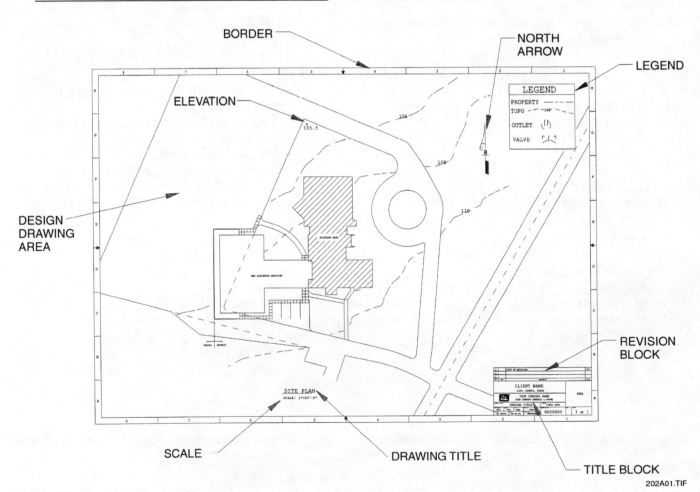

Practice Problem I ◆ The site plan.

Practice Problem II:
The Sanitary Plumbing Plan

The illustration for *Practice Problem II* shows a detail of the sanitary plumbing plan. Use this plan to answer the following questions.

1. Where is the chase of the plumbing?

    _____
    _____
    _____
    _____

2. Give the direction of the walls (north, east, south, west) where the plumbing fixtures are located for:

    the men's room   _____

    the women's room   _____

3. What is the dimension of the pipe that extends directly to the sewer?

    _____

Practice Problem III:
Fixtures and Plumbing System

Refer to the figures on the following page (*Figures 5* and *6* from the text) to answer these questions.

1. Where would you find the list of fixtures for this addition?

    _____
    _____
    _____
    _____
    _____
    _____
    _____

2. Where would you find the sizes of the DWV and water supply lines?

    _____
    _____
    _____
    _____
    _____
    _____

3. Where would you find directions for how to test the plumbing system?

    _____
    _____
    _____
    _____
    _____
    _____

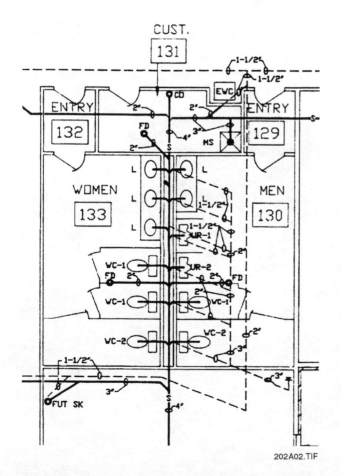

202A02.TIF

Practice Problem II ◆ Plumbing plan detail.

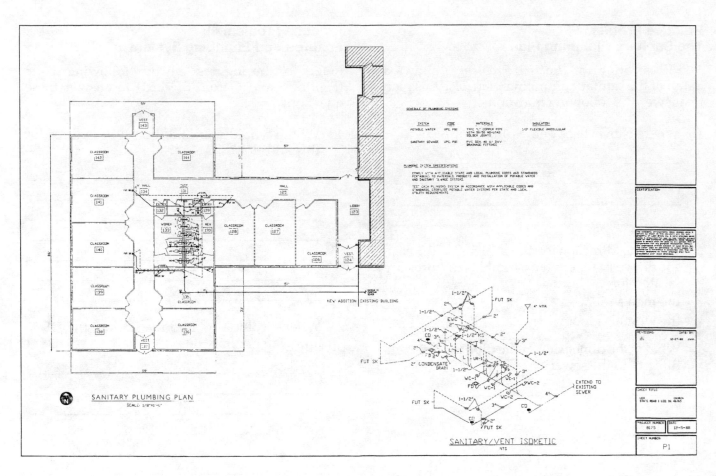

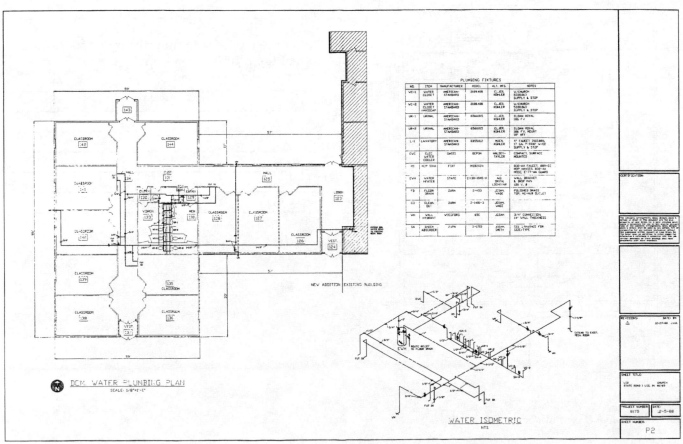

Trade Terms Introduced in This Module

Alternate: In commercial drawings, a term that describes other possible versions of the basic project with items either added or removed as specified by the owner.

Architectural drawing: A drawing that shows the shape, size, and appearance of a structure.

Chase: A continuous recess built into a wall to receive pipes, ducts, and so on. In commercial plumbing, a hollow wall built between two back-to-back restrooms to hold piping.

Civil drawing: A drawing that shows the overall shape of the building site. Also called the *site plan*.

Cut sheet: A drawing of a specific fixture with rough-in dimensions and other information.

Electrical drawing: A drawing that shows the details of a building's electrical system.

Elevation: A drawing that shows the vertical elements of a building, either inside or outside.

Floor plan: A drawing that shows the size and shape of the floor of a building.

Isometric drawing: A three-dimensional projection in which all of the principal planes are drawn parallel to corresponding established axes and at true dimensions.

Landscape drawing: A drawing that shows proposed plantings and other landscape features.

Mechanical drawing: A drawing that shows the location and size of the components of the heating, ventilating, and air-conditioning (HVAC) system.

North arrow: In commercial drawings, an arrow that indicates where north is relative to the site plan. Various trades use the arrow as an orientation point to locate building components properly.

Plumbing drawing: A drawing that shows the location of the drain, waste, and vent (DWV) and water supply piping.

Request for Information (RFI): A form used to question discrepancies on the drawings or to ask for clarification.

Riser diagram: A diagram that shows how the piping system is to be installed; frequently included on the plumbing drawings.

Schedule: A detailed list of components, items, or parts to be furnished in a building project—for example, door and window schedules.

Site plan: See civil drawing.

Structural drawing: A drawing that shows how the framework of the building is to be constructed.

Title block: In commercial drawings or blueprints, the block that contains the drawing's scale, the date of last revision, the drawing number, and the name of the architect or engineer.

Piping Symbols

WASTE WATER

DRAIN OR WASTE-ABOVE GRADE	——————
DRAIN OR WASTE-BELOW GRADE	— — — ·
VENT	- - - - - - -
COMBINATION WASTE AND VENT	—CWV—
ACID WASTE	—AW—
ACID VENT	- - - AV - - -
INDIRECT DRAIN	—D—
STORM DRAIN	—SD—
SEWER-CAST IRON	S-CI
SEWER-CLAY TILE BELL & SPIGOT	S-CT
DRAIN-CLAY TILE BELL & SPIGOT	——————

OTHER PIPING

GAS-LOW PRESSURE	—G—G—
GAS- MEDIUM PRESSURE	—MG—
GAS-HIGH PRESSURE	—HG—
COMPRESSED AIR	—A—
VACUUM	—V—
VACUUM CLEANING	—VC—
OXYGEN	—O—
LIQUID OXYGEN	—LOX—

202A03.TIF

Material Takeoff

Materials Estimate						

Project:				Estimate No.		
Location:				Sheet No.		
Architect/Engineer:				Date:		
Summary by:		Prices by:		Checked by:		

Quantity	Description	Unit Price Material Cost	Total Estimated	Unit Price Labor Cost	Total Estimated	Total
	TOTAL COSTS					

Answers to Review Questions

Answer	Section
1. a	2.1.0
2. c	2.1.0
3. a	2.3.0
4. d	2.4.0
5. c	2.4.0
6. b	3.0.0
7. c	4.0.0
8. a	4.1.0
9. d	4.2.0
10. d	4.3.0

Answers to Practice Problems

Practice Problem I

1. The scale is 1" = 20'-0".

2. In the title block.

3. The hatched lines or shaded area show the existing building. In addition, labels on the drawing show which is the existing building and which is the addition.

4. The west wall.

5. The broken dashed lines show the property line as indicated by the legend in the upper right corner of the drawing.

Practice Problem II

1. The chase runs between the men's restroom and the women's restroom.

2. The west wall of the men's restroom; the east wall of the women's restroom.

3. The dimension of the pipe is 4".

Practice Problem III

1. On *Figure 6* under the heading Plumbing Fixtures.

2. On *Figure 6* above the heading Water Isometric.

3. On *Figure 5* under the heading Plumbing System Specifications.

NCCER CRAFT TRAINING USER UPDATES

The NCCER makes every effort to keep these textbooks up-to-date and free of technical errors. We appreciate your help in this process. If you have an idea for improving this textbook, or if you find an error, a typographical mistake, or an inaccuracy in the NCCER's Craft Training textbooks, please write us, using this form or a photocopy. Be sure to include the exact module number, page number, a detailed description, and the correction, if applicable. Your input will be brought to the attention of the Technical Review Committee. Thank you for your assistance.

Instructors – If you found that additional materials were necessary in order to teach this module effectively, please let us know so that we may include them in the Equipment and Materials list in the Instructor's Guide.

Write: Curriculum Revision and Development Department
National Center for Construction Education and Research
P.O. Box 141104, Gainesville, FL 32614-1104

Fax: 352-334-0932

E-mail: curriculum@nccer.org

Craft _____   Module Name _____

Copyright Date _____   Module Number _____   Page Number(s) _____

Description _____

_____

_____

_____

(Optional) Correction _____

_____

_____

(Optional) Your Name and Address _____

_____

_____

Installing and Testing DWV Piping

COURSE MAP

This course map shows all of the modules in the second level of the Plumbing curriculum. The suggested training order begins at the bottom and proceeds up. Skill levels increase as you advance on the course map. The local Training Program Sponsor may adjust the training order.

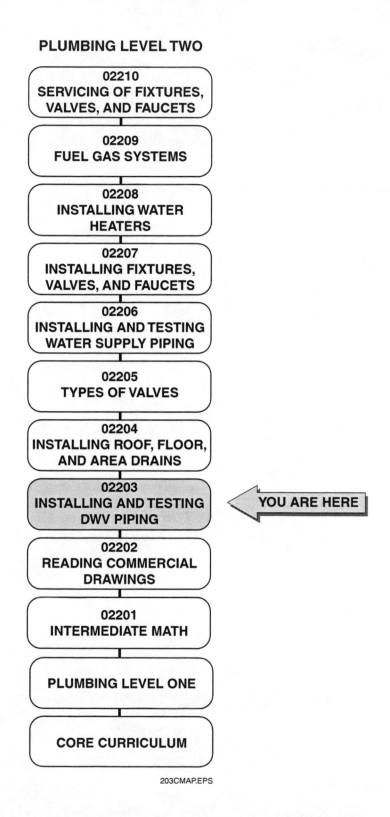

PLUMBING LEVEL TWO

02210
SERVICING OF FIXTURES, VALVES, AND FAUCETS

02209
FUEL GAS SYSTEMS

02208
INSTALLING WATER HEATERS

02207
INSTALLING FIXTURES, VALVES, AND FAUCETS

02206
INSTALLING AND TESTING WATER SUPPLY PIPING

02205
TYPES OF VALVES

02204
INSTALLING ROOF, FLOOR, AND AREA DRAINS

02203
INSTALLING AND TESTING DWV PIPING ⟵ YOU ARE HERE

02202
READING COMMERCIAL DRAWINGS

02201
INTERMEDIATE MATH

PLUMBING LEVEL ONE

CORE CURRICULUM

203CMAP.EPS

MODULE 02203 CONTENTS

Figures

Tables

Installing and Testing
DWV Piping

Objectives

Upon completion of this module, you will be able to do the following:

1. Develop a material takeoff from a given set of plans.
2. Use plans and fixture rough-in sheets to determine location of fixtures and route of the plumbing.
3. Demonstrate the ability to install a building sewer.
4. Locate the stack within the structure.
5. Demonstrate the ability to install a DWV system using appropriate hangers and correct grade.
6. Demonstrate the ability to modify structural members using the appropriate tools without weakening the structure.
7. Demonstrate the ability to test a DWV system.

Prerequisites

Before you begin this module, it is recommended that you successfully complete the following modules: Core Curriculum; Plumbing Level One; Plumbing Level Two, Modules 02201 and 02202.

Required Trainee Materials

1. Paper and sharpened pencil
2. Appropriate Personal Protective Equipment

1.0.0 ◆ INTRODUCTION

Drain, waste, and vent (DWV) piping is installed before the water supply piping because it is less easily handled, it must maintain a specific **slope**

(grade), and it is easier to route water piping around the DWV piping than vice versa.

The complexity of a DWV installation can vary widely, but a good practice for any installation is an action plan that outlines how the installation will proceed. A good plan will include factors such as the following:

- Depth of the inlet to the sewer or private sewage system
- Finished floor elevations where fixtures are installed
- Calculations of the developed length of DWV piping versus the requirement for slope
- Requirements for excavation including safety considerations like trench **shoring**
- Identification of the main structural elements such as foundation footers, concrete slabs, floor systems, roof systems, and stairs
- Calculations of drain and vent sizing based on the load requirements
- Identification of the plans and specifications to be used
- Types of fixtures and fixture supports and their locations within the building
- Coordination of measurements and timing with other trades
- Delivery of **time-critical material**, such as a one-piece fiberglass tub/shower unit
- Delivery and storage of materials
- Arrangements for inspection and equipment for testing
- **Accessibility requirements**

How the installation proceeds depends on how the building is being built. In **slab-on-grade** construction, you have to start work before the walls are constructed. On the other hand, a building may be totally framed up, lacking only the basement slab and the roof. In this case, you can complete the entire installation at one time. How you order materials, use equipment, and schedule inspections are all affected by how the building is being built, and your action plan should reflect this.

2.0.0 ◆ PLANS

For economy and efficiency, plumbers work according to plans or blueprints. For most industrial, commercial, and multi-residential structures, plumbing plans are included with the architectural plans for the structure. These plans are designed with a certain amount of flexibility built in to allow the plumber to make on-the-job decisions when confronted by unusual circumstances. In many cases, particularly in light residential construction, plumbing plans are not included with the prints. In these cases, the plumber uses the floor plan to locate the runs of pipe and the various plumbing fixtures (see *Figure 1*).

2.1.0 Material Takeoffs

A **material takeoff** is a list of the types and quantities of materials required for a given job. An estimator who is knowledgeable about plumbing practices and materials may work from a set of drawings to do the takeoff. There is no standard form used for material takeoffs. A typical form is shown in *Figure 2*.

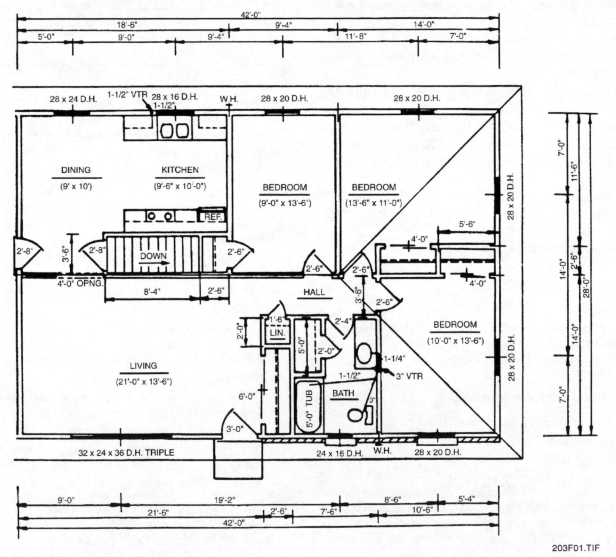

Figure 2 ◆ Material takeoff with estimate.

Figure 1 ◆ House floor plan.

2.2.0 Locating Plumbing Fixtures

Locating the fixtures according to the floor plan is the first step in plumbing a structure. The location of the DWV stacks and the route of the fixture branch lines to the stack can then be established from the plumbing details. If plumbing details are not available, a plumber relies on practical experience to determine stack locations and branch line routes to the stacks. Throughout this module, we will use a single-story house with a basement as an example.

2.2.1 Rough-In Measurements

After determining the approximate location of the fixture from the floor plan, study the rough-in measurements (see *Figure 3*). You will find these measurements in manufacturers' catalogs, which are available from plumbing distributors. Note that critical dimensions may vary between styles of the same brand (refer to A and B in *Figure 3*). Rough-in measurements tell you exactly where the piping should exit the walls or floors. You will use these dimensions for reference later

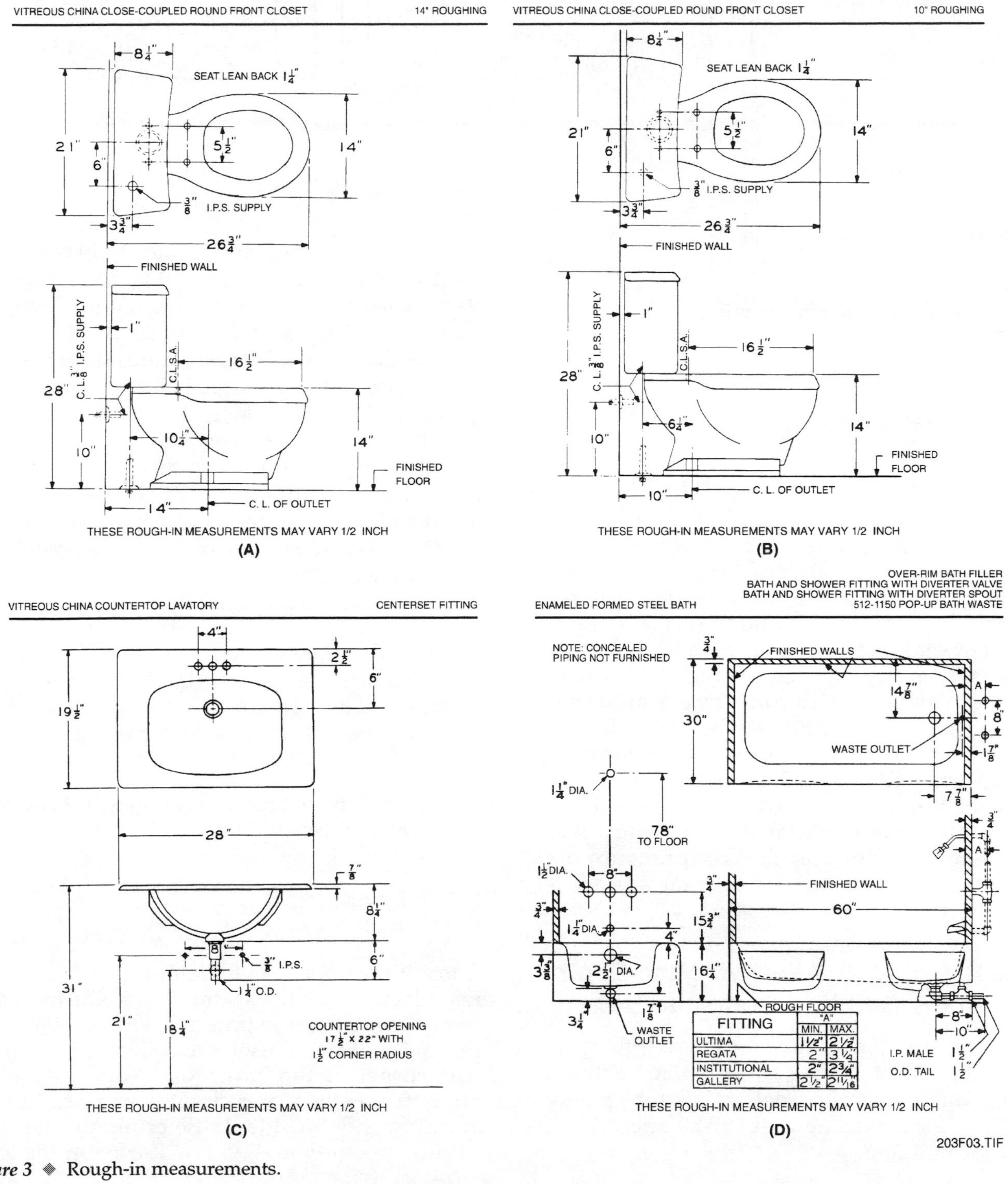

Figure 3 ◆ Rough-in measurements.

to run the water supply piping and to attach the fixture to the floor or wall. When determining where to locate the holes for the DWV piping, you must consider the thickness of the finished wall or floor over which the fixture will be attached (see *Figure 4*). Add these dimensions to those of the rough-in measurements to ensure accuracy.

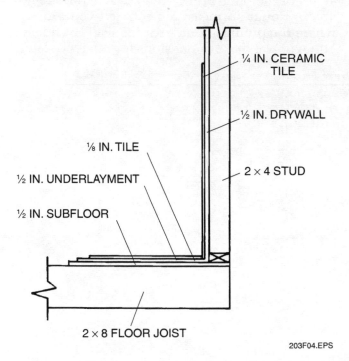

Figure 4 ◆ Thickness of wall-finishing materials.

Now you must determine the required fixture support. Are special **carrier fittings** being used? You must consider the dimensions of these devices and note the locations of the components. For example, if you install a lavatory carrier (see *Figure 5*) that requires a 3½-inch stud wall, will you be able to run a trap arm behind the vertical portions of this carrier? The width of the vertical portion plus the outside diameter of the trap arm will have to be less than 3½ inches. If it is not, you will have to install an additional stack directly behind the lavatory to avoid a conflict. Coordination of the fixture support elements of any installation is vital to a well-planned job.

2.3.0 Locating and Verifying the Depth of the Building Drain and Sewer

You must verify the location and the depth of the **sewer tap**, or inlet, of the private sewage disposal system. This is critical for determining the piping route and for determining its installation depth inside the building.

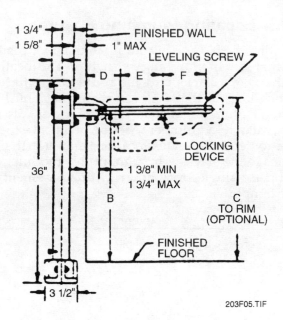

Figure 5 ◆ Lavatory carrier.

For example, a half-bath (toilet and lavatory) is located in a basement. The fittings that must be stacked beneath the floor are a long-turn 90-degree ell and a sanitary tee with a closet bend attached. This combination of fittings requires 16 inches from the bottom of the 90-degree ell to the top of the closet bend (finished floor). The length of the run from this stack to the sewer inlet is 100 feet. In this case, 3-inch pipe is used, so this pipe requires at least a ¼-inch-per-foot slope. To calculate the depth this run of pipe requires, multiply the slope by the length of the run. This run of pipe would require 25 inches of depth.

¼ inch per foot × 100 feet = 25 inches

Add the required amount of slope to the height of the fittings used below the lowest floor with plumbing. The minimum depth required for the sewer tap from the finished floor of the basement is 41 inches.

16 inches (the combination of fittings) + 25 inches (depth) = 41 inches

2.4.0 Locating the Residential Water Closet

Consult the floor plan to determine the approximate location of the water closet. On the bathroom floor plan illustrated in *Figure 6*, you can see that the water closet is to be located in the left rear corner of the bathroom. You must accurately center the water closet from front to back and from side to side. In determining the side-to-side measurement, you will consult the local plumbing code, and sometimes, the building code. The code will state the minimum permissible distance between the centerline of the water closet and a counter, shelf, partition, or in this case, a wall. Let's assume that the minimum permissible distance from the finished wall (on each side) to the centerline of the water closet is 15 inches.

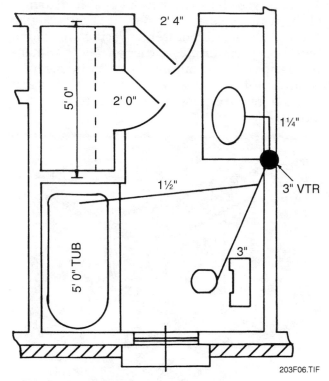

Figure 6 ◆ Bathroom floor plan.

To accurately locate the water closet, you should also consider the thickness of the wall-finishing materials. If the finished wall is to consist of ¼-inch ceramic tile over ½-inch drywall, add ¾ inch to the minimum permissible distance, making the centerline of the waste outlet from the rough framing 15¾ inches (refer to *Figure 4*). You would make a mark parallel to the sidewall at 15¾ inches.

Consult the manufacturer's rough-in measurements to determine the distance from the rear wall to the center of the closet flange. This will center the water closet from front to back. Referring to *Figure 3 (A)*, you'll see that this distance is 14 inches, and that it includes a 1-inch clearance space at the rear. Because the wall is rough-framed, you must add the thickness of the wall-finishing material to this dimension. If ½-inch drywall covered with ¼-inch ceramic tile were to be used, you would add ¾ inch, making a total of 14¾ inches from the framed rear wall to the centerline of the closet flange. Mark this measurement on the floor, perpendicular to the line already laid out (see *Figure 7*). The two crossed

lines represent the exact center of the location of the closet flange. Cut the hole for the closet flange with a drill and scroll saw.

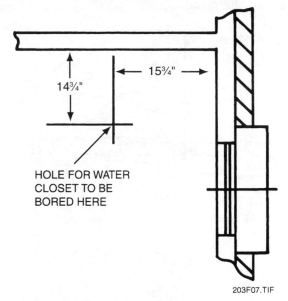

Figure 7 ◆ Laying out the location of the water closet.

2.5.0 Locating the Commercial Water Closet

When roughing-in commercial fixtures, you must consider the possibility of abuse and vandalism. Greater structural strength and resistance to both accidental and deliberate abuse are necessary and involve the use of carrier fittings. Additional considerations include ease of maintenance, ease of installation, cost, and appearance. Specific installation requirements and rough-in dimensions depend entirely on the make and model of the carriers that are being used. You should have the **manufacturers' specification sheets** on hand before attempting to lay out the piping to these devices.

Wall-hung water closets require special mounts and special DWV piping, which can result in a more difficult installation and greater cost. However, wall-hung closets offer advantages such as easier cleaning and maintenance. The open area below the stool makes regular maintenance of the restroom floor easier (see *Figure 8*). In addition, wall-hung closets offer savings in other areas. Since the drainage from the water closet is through the wall, no holes are necessary in the floor at each closet location. When dealing with concrete floors, this can be a substantial labor saver. Holes that normally would be required for the vent stack also are eliminated and the false ceiling, used to conceal pipes, is no longer necessary. These factors all contribute to savings in time and labor.

A typical carrier for a wall-hung water closet is shown in *Figure 9*. These carriers are available in both horizontal and vertical outlet models for different applications.

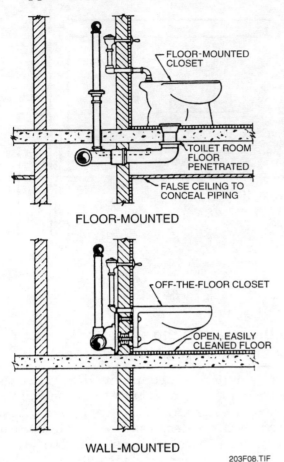

Figure 8 ◆ Floor- and wall-mounted water closets.

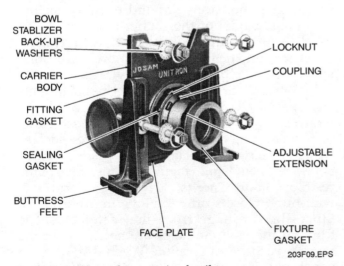

Figure 9 ◆ Water closet carrier detail.

A typical installation showing back-to-back restrooms with a plumbing access area is shown in *Figure 10*, which also illustrates how the components function together as a unit. Double-system units for back-to-back installations are used along with lateral pipe connections to span the width of the access area between restrooms. Auxiliary faceplates are applied to the lateral members to hook up the fixtures in the adjacent restrooms. A plan view of these components and the vent piping can be seen in *Figure 11*.

The plumbing access area between two restrooms (the **chase**) may be a variable distance, but the wall thickness has definite limits. *Figure 12* shows one manufacturer's specifications for a back-to-back water closet carrier. These dimensions range from a minimum of 2¼ inches to a maximum of 6¼ inches on this model. Several other ranges are possible on different carrier models.

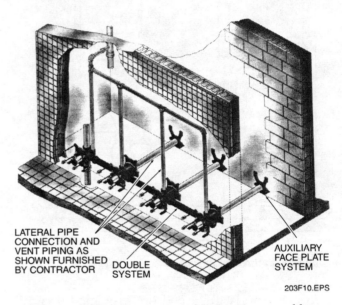

Figure 10 ◆ Back-to-back water closet carrier assembly.

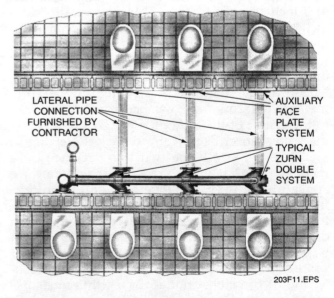

Figure 11 ◆ Plan view of a back-to-back water closet carrier assembly.

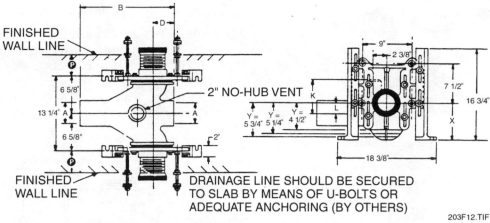

Figure 12 ◆ Back-to-back water closet carrier.

Another possibility that exists with some carrier models is the invertible faceplate shown in *Figure 13*. This feature allows for level placement of the water closets while maintaining a pitch in the drainage line. The maximum difference in the elevation of the fixture port is 4¼ inches.

An engineer will provide necessary specifications for installations requiring a battery of wall-hung water closets. Careful review of the working drawings and equally careful installation practices will ensure a properly functioning assembly.

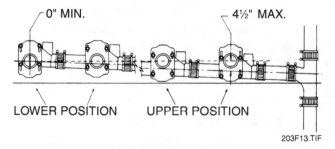

Figure 13 ◆ Invertible faceplate.

2.6.0 Locating Urinals and Other Fixtures

When locating urinals and other fixtures such as water coolers, be aware that the rough-in dimensions are totally dependent on the make of the fixture being used. You will have less room for error on the rough-in on these types of fixtures, so you will need the manufacturer's specification sheets on hand.

2.6.1 Urinals

The carriers most often used for installing urinals are similar to those used for lavatories (see *Figure 14*). Variations occur in the cross member, which is designed to accept bolts from the fixture itself. A second cross member is used to mount the base of the urinal. Models also are available with drainage ports and in an alternate design, shown in *Figure 15*.

Figure 14 ◆ Double-plate, floor-mounted urinal carrier.

203F15.EPS

Figure 15 ◆ Blowout urinal carrier with adjustable coupling.

An additional modification available for urinal carriers is a mount for the water supply piping. This mount provides the correct water supply location for proper installation of the flush valve. It also holds the water supply piping firmly in place to prevent damage to the supply piping or the valve.

In addition to the fixture carriers mentioned, other carrier types include perineal bath, water cooler, and corner sink varieties. Perineal bath fixtures are found primarily in hospitals and require a special carrier for wall-hung units (see *Figure 16*). These carriers are floor-mounted and are equipped with a second hanger for the top portion of the fixture.

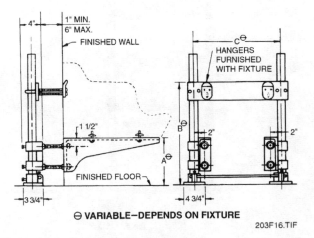

203F16.TIF

Figure 16 ◆ Perineal bath carrier.

2.6.2 Water Coolers

Carriers are also available for water coolers (see *Figure 17*). Variable widths are manufactured for the many varieties of coolers available. Two hangers are required to properly secure the water cooler to the wall. Side-by-side and back-to-back carriers are also available for water coolers.

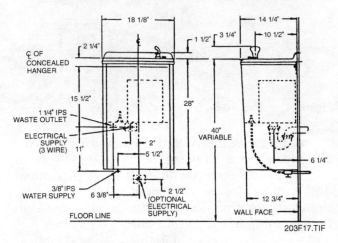

203F17.TIF

Figure 17 ◆ Wall-mounted water cooler.

2.6.3 Corner Sinks

Special carriers can be purchased for corner sinks, which provide the secure structural support required for these fixtures. Most fixtures available for commercial installation can be wall mounted to provide ease of cleaning and maintenance. Be sure that the carriers you order are compatible with the fixtures planned for the installation.

2.7.0 Locating the Residential Lavatory

Refer to the bathroom floor plan in *Figure 6* to determine the approximate location of the lavatory. You can see that the lavatory will be located in the left front part of the bathroom. The floor plan also shows that the lavatory will be inserted into the right side of a bathroom countertop. So you must determine the location of the lavatory within the countertop.

First, determine the overall width and length of the lavatory. You can get this information by referring to the rough-in measurements in *Figure 3(C)*. Measure the width and depth of the countertop. Ideally, the lavatory should be installed with adequate front, back, and side clearance. This information is rarely provided on the blueprints. Plumbers generally rely on experience to determine proper placement of the lavatory.

Next, mark the point of exit of the drainpipe. The rough-in measurements indicate that the drain is centered in the lavatory. That center is the point to be marked. However, because the structure is only roughed-in, the countertop is not yet in position. Plumbers usually place a mark on the **soleplate** to indicate the center of the lavatory drain to show where it will pass inside the framed wall.

3.0.0 ◆ BASIC FRAMING FOR LAVATORIES AND SINKS

Basic framing in residential and light commercial structures usually accommodates plumbing installations without any unusual problems. An enlarged plumbing wall can be built to house the piping, and various forms of **blocking** can be installed, which will later serve as anchors for screws that will hold lavatory brackets (see *Figure 18*). These fixture rough-in procedures are the most basic and most often used.

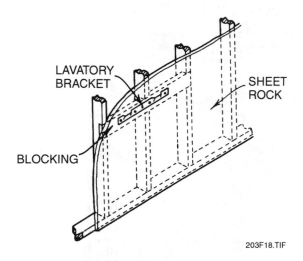

203F18.TIF

Figure 18 ◆ Wall section showing lavatory bracket.

3.1.0 Locating Commercial Lavatories and Sinks

The more common method of supporting wall lavatories and sinks in commercial structures is to use metal support carriers concealed within the wall (see *Figure 19*). Carriers are adjustable and are available with arms that are concealed or exposed.

Leveling screws, which allow the fixture to be leveled on an uneven floor, are provided with most commercial carriers. One manufacturer provides adjustable block base feet for this purpose (see *Figure 20*). Note that these must be anchored securely.

Lavatory carriers also may be mounted directly to wall surfaces. These carriers do not use vertical uprights but are equipped with metal plates for mounting on wood or metal stud walls or masonry walls (see *Figure 21*). These carriers are convenient but have poor structural strength.

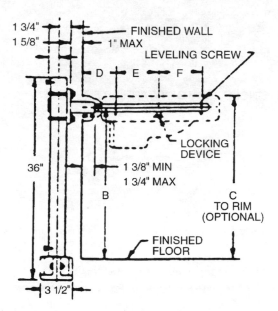

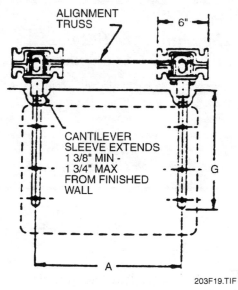

203F19.TIF

Figure 19 ◆ Lavatory carrier.

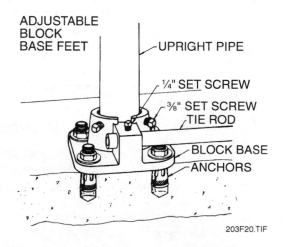

203F20.TIF

Figure 20 ◆ Block base feet.

(A) SINGLE LAVATORY CARRIER FOR STUD WALL

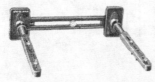

(B) SINGLE LAVATORY CARRIER FOR MASONRY WALL

203F21.EPS

Figure 21 ◆ Single lavatory carriers for stud and masonry walls.

Lavatory carriers are available for back-to-back installations (see *Figure 22*). Because back-to-back lavatory installations share the same hardware, you can realize significant savings in costs and time. The headers (arm mounts) on each side of the carrier are designed to mesh so that lavatory heights can be equal in both rooms.

Various kinds of sink carriers are available for a wide variety of sink designs and applications. The primary difference between lavatory carriers and sink carriers is their structural strength.

Because the sink is generally heavier than the lavatory and can hold more water, additional structural considerations are necessary. For this reason, sink carriers are of the floor-mounted variety to maximize the carrying capacity by transferring the weight directly to the floor.

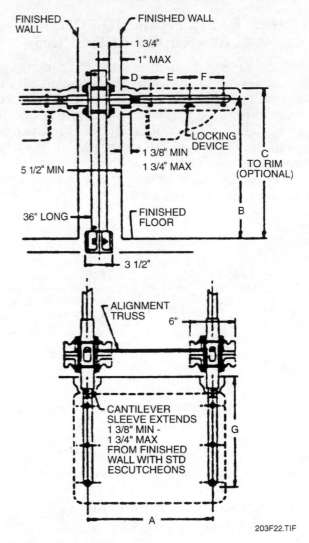

Figure 22 ◆ Two-wall lavatory carrier.

A floor-mounted sink carrier with a waste line bracket included as part of the carrier assembly is shown in *Figure 23*. These carriers provide support for the sink as well as for the drain line. This component eliminates the need to construct additional supports for the DWV piping.

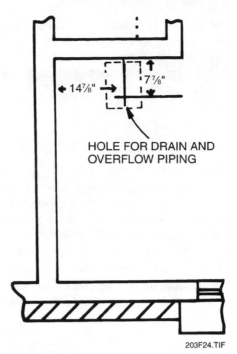

Figure 24 ◆ Locating the bathtub drain.

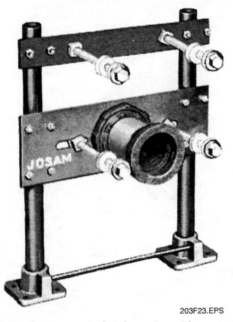

Figure 23 ◆ Floor-mounted sink carrier with waste fitting.

3.2.0 Locating the Bathtub

Refer to the bathroom floor plan in *Figure 6*. You'll see that the bathtub is located on the wall opposite the water closet. Refer to the rough-in measurements in *Figure 3 (D)*. You'll see that the bathtub drain is located 14⅞ inches from its left side wall. Because of their size, bathtubs are fitted into place before the finished walls are up. The finished wall will rest on the upper rim of the bathtub. Therefore, the rough-in measurements are made from the rough-framed wall. The 14⅞-inch measurement is laid out from the wall on the left side and is marked on the floor (see *Figure 24*).

Consulting the rough-in measurements again, you'll see that the bathtub drain is located 7⅞ inches from the rough-framed wall at the head of the bathtub. This measurement is also laid out on the floor. The crossed lines represent the center of the bathtub drain.

Before cutting the drain hole in the floor, consider the distance from the bathtub overflow pipe to the bathtub drain. The hole must be large enough to provide clearance for both of these. Also, you will need to provide clearance so that you'll have enough room to use pipe wrenches to make up the joints. Typically, you would cut a hole approximately 12 inches long and 8 to 12 inches wide.

3.2.1 Tub and Shower Rough-In

Give special attention to one-piece tub and shower fixtures made of fiberglass. These fixtures are too large to be installed after framing is completed. They must be in place during construction. Take care to properly frame the opening for these fixtures, as the dimensions are critical. Fixture rough-in measurements do not include an interior wall covering such as **sheetrock** and must be exact from stud wall to stud wall. Often, additional 2 × 4 framing is required to support the flexible upper portions of the enclosure. This additional framing must be in place before the sheetrock is applied to the opposite side of the wall. Although most of these considerations are the carpenter's responsibility, a good plumber will oversee the rough-in work to be sure that the job is done correctly.

Keep the carton in which the fiberglass tub or shower enclosure was shipped. Use it to protect the fixture from sunlight, weather, and damage that could be caused by other trades during construction.

3.3.0 Locating Fixtures for the Physically Challenged

Special considerations apply when roughing-in fixture installations for physically challenged individuals. For example, wall-mounted water closets for the physically challenged are the same as standard water closets. The difference is the height of the opening and mounting bolts on the carrier fitting. In a horizontal installation of many carrier fittings (a battery installation), the toilet for the physically challenged must be at the upper end of the piping run to accommodate the higher rough-in of the special fixture.

Figure 25 illustrates a typical bathroom for the physically challenged. Note that the room's dimensions include a 5-foot turning circle so that a wheelchair can turn easily. The final dimensions required are illustrated for both fixtures. You must adjust fixture rough-ins so that they comply with these dimensions.

Lavatories and water cooler carriers are available for wheelchair access. The fixtures themselves have shallow features, and the support arms from the carriers are hidden and extend out farther to give support to the fixture.

Be sure to refer to the American National Standards Institute (ANSI) code and your local code for provisions that apply to physically challenged individuals.

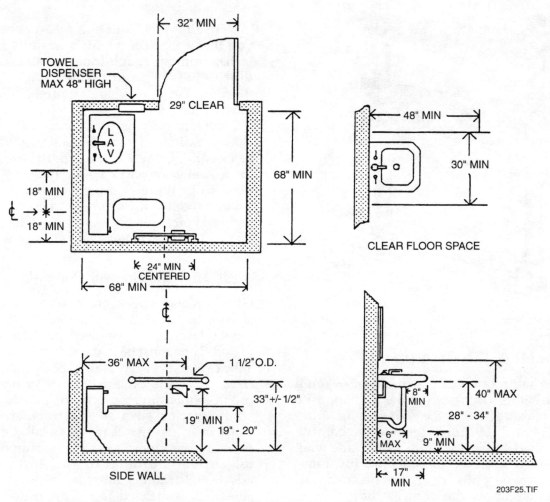

Figure 25 ◆ Fixture installation for the physically challenged.

ON THE
· LEVEL ·

ANSI Standards
American National Standard Institute—
Accessible and Usable Buildings and Facilities

1   Purpose and Application
1.1   Purpose

The specifications in this standard make buildings and facilities accessible to and usable by people with such physical disabilities as: the inability to walk, difficulty walking, reliance on walking aids, blindness and visual impairment, deafness and hearing impairment, incoordination, reaching and manipulation disabilities, lack of stamina, difficulty interpreting and reacting to sensory information, and extremes of physical size based generally upon adult dimensions. Accessibility and usability allow a person with a physical disability to independently get to, enter, and use a building or facility.

This standard provides specifications for elements that are used in making functional spaces accessible. For example, it specifies technical requirements for making doors, routes, seating, plumbing fixtures, and other elements accessible. These accessible elements are used to design accessible functional spaces such as classrooms, bathrooms, hotel rooms, lobbies, or offices.

This standard is for adoption by government agencies and by organizations setting model codes to achieve uniformity in the technical design criteria in building codes and other regulations. This standard is also used by non-governmental parties as technical design guidelines or requirements to make buildings and facilities accessible to and usable by persons with physical disabilities.

Review Questions

Sections 2.0.0–3.3.0

1. An estimator lists the type and quantity of plumbing supplies needed for a job on the _____.
   a. architect's blueprint
   b. material takeoff
   c. elevation drawing
   d. building plan

2. The approximate location of fixtures usually is determined by the _____.
   a. number of fixtures in the building
   b. manufacturer's catalog
   c. local building code
   d. floor plan

3. A residence with a basement half-bath requires a 90-degree ell and a sanitary tee that need 16 inches of depth below the finished floor. The sewer inlet is 60 feet from the stack, and a slope of at least ¼ inch per foot is required. The minimum depth of the sewer tap below the finished basement floor is _____.
   a. 15 inches
   b. 16 inches
   c. 31 inches
   d. 41 inches

4. You must accurately center a water closet _____.
   a. from front to back
   b. from side to side
   c. from front to back and side to side
   d. along the centerline of the main building stack

5. Be sure to refer to the _____ for provisions that apply to physically challenged individuals.
   a. ANSI standards and the local code
   b. ANSI and OSHA standards
   c. Uniform Plumbing Code
   d. manufacturer's rough-in sheets

4.0.0 ◆ DETERMINING THE LOCATION OF THE STACK

As the greatest amount of discharge in a home comes from the bathroom fixture group, the main stack is located within one of the bathroom walls. The stack generally is located near the water closet, which has the largest discharge rate of all common household plumbing fixtures. In determining the exact location of the stack, consider the following:

- Direction and path of pipe from the fixtures
- Ease in plumbing
- Economy

Refer to *Figure 6*, which illustrates the placement of the DWV stack in the bathroom floor plan. Mark the location of the stack on the center of the soleplate. You must center the stack in the wall to prevent the pipe from accidentally being struck by nails and other fasteners. This also allows more flexibility in the placement of fittings within the wall. Cut an appropriately sized hole in the soleplate with a hole saw, multi-spur bit, or saber saw.

4.1.0 Centering the Stack Within the Building

After you locate the stack on the soleplate and bore a hole, you can determine the location of the fitting at the base of the stack and the stack exit through the roof. The stack must run true vertical. To ensure this, use a plumb bob and string (see *Figure 26*).

Suspend the plumb bob from the center of the hole you cut in the floor for the stack. After the plumb bob has stopped moving, it will indicate the location of the center of the sweep at the base of the stack (see *Figure 27*).

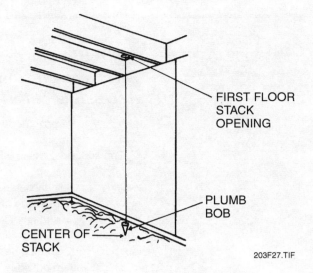

FIRST FLOOR STACK OPENING

PLUMB BOB

CENTER OF STACK

203F27.TIF

Figure 27 ◆ Locating the centerline of the sweep.

Use the plumb bob and string to determine where the stack will exit the roof. To do this, first cut a hole in the **top plate**. Hold the string against the top plate and move it around until the plumb bob comes to rest in the center of the hole cut in the soleplate (see *Figure 28*). Once the plumb bob stops moving, the point where the string touches the top plate marks the center of the path the pipe will take through the top plate. Using that center, bore an appropriately sized hole for the stack to pass through.

203F26.EPS

Figure 26 ◆ Plumb bob.

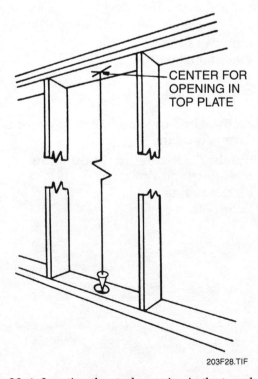

CENTER FOR OPENING IN TOP PLATE

203F28.TIF

Figure 28 ◆ Locating the stack opening in the top plate.

To determine where the stack will exit the roof, hold the string against the roof **sheathing**, while the plumb bob hangs through the hole in the top plate, to the hole in the soleplate. Move the string around until the plumb bob comes to rest in the center of the hole cut in the soleplate (see *Figure 29*). The point where the string touches the roof marks the exit of the stack through the roof. Mark this location and cut an appropriately sized hole in the roof.

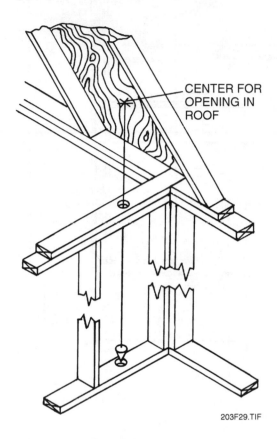

CENTER FOR OPENING IN ROOF

203F29.TIF

Figure 29 ◆ Locating the stack exit through the roof.

4.2.0 Installing the Building Drain

After you have located the stack, begin work on the building drain. By running the building drain before installing the stack, you will ensure that the sweep at the base of the stack is located at the proper elevation. If the sweep is not accurately located, wastes will not flow properly from the stack to the building drain.

The building drain may be run according to plumbing plans. However, in cases where plumbing plans are not provided, you must determine the route of the building drain. In determining this path, consider the location of the floor drain, stacks, and fixture drains. In our example, the building drain will begin at the secondary stack, as it is the furthest drain (see *Figure 30*). From there, it will join with the floor drain and the primary stack. At the primary stack, the building

drain will run under the foundation footing and then continue to the sewer main. Therefore, you must locate the point where the building drain passes under the foundation footing and the centers of the two stacks before beginning to run the building drain.

4.3.0 Locating the Centerline of the Main Stack

Start by locating the center of the sweep at the base of the main stack using a plumb bob and string. Suspend the plumb bob through the center of the hole bored in the soleplate for the stack. Lower the plumb bob until it is approximately ¼ inch off the ground. This point represents the location of the center of the main stack at the level of the unfinished floor.

4.4.0 Locating the Building Drain

After locating the centerline of the main stack, you can determine where the building drain will pass under the foundation footing. For economy, the building drain usually runs straight to the sewer main. In our example, this means that the building drain would run perpendicular to the foundation wall (cross it at 90 degrees) and continue to the sewer main. Lay out a line from the centerline of the main stack perpendicular to the foundation wall. This is the path the building drain will take on its way to the sewer main.

4.5.0 Locating the Centerline of the Secondary Stack

Next, determine the location of the secondary stack by suspending a plumb bob and string through the hole you cut for the stack. Lower the plumb bob until it is suspended approximately ¼ inch above the ground. Mark that point on the ground.

 DID YOU KNOW?

One-Piece Bathtub with Apron

In 1911, the Kohler Company introduced the one-piece bathtub with apron. Its development was considered revolutionary in the industry. Before this change, built-in bathtubs were cast in two sections: the tub and the apron. The apron is the section of the tub that runs from the rim to the floor. The plumber fitted these two pieces together during installation. The one-piece tub, which eliminated all joints and seams, was more sanitary and attractive.

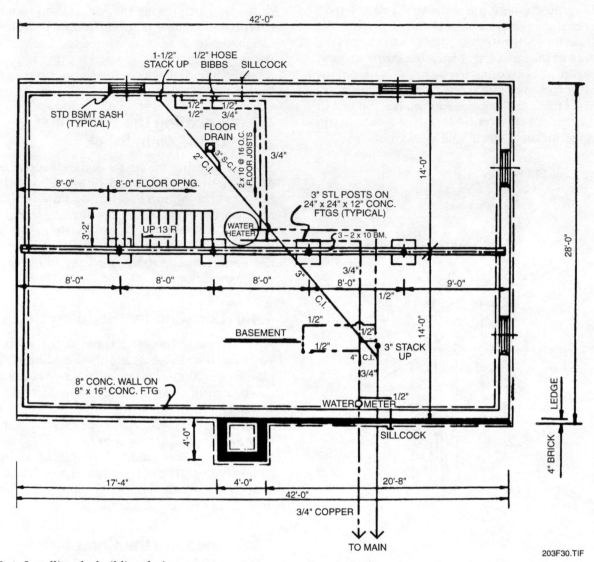

Figure 30 ◆ Installing the building drain.

203F30.TIF

4.6.0 Calculating the Depth of the Building Drain

After locating the major points along the building drain, calculate the slope and depth of the pipe. You must determine the depth at which the building drain will pass under the foundation footing. This information may be found on the plumbing plans. If it is not, you will have to dig a hole beside the foundation wall and measure the depth to the underside of the footing.

To this measurement, add the desired distance between the centerline of the pipe and the underside of the foundation. For example, suppose the foundation footing is 22 inches below the finished level of the floor. Also suppose that you want the centerline of the pipe to pass 8 inches under the footing. Therefore, the depth of the centerline of the building drain is 30 inches below the finished floor level (22" + 8" = 30").

4.7.0 Calculating the Elevation of the Main Stack Sweep

After calculating the depth of the building drain at the foundation wall, you can determine its rise to the sweep at the base of the main stack. The drain must rise for the stack to drain into it. You can do this by referring to the plans or by direct measurement. Let's say that the sweep is 6 feet from the foundation wall. At a slope of ¼ inch per foot, the building drain would rise 1½ inches ($6 \times$ ¼" = 1½").

After determining the fall, you can figure out the elevation of the sweep. In this example, you know that the building drain centerline will pass under the foundation wall 30 inches below the finished floor level. You also know that the drain will rise 1½ inches at the centerline of the sweep. Therefore, the centerline of the sweep will be located 28½ inches below the level of the finished floor (30" – 1½" = 28½").

Because these measurements are made to the centerline of the pipe, you must consider the pipe's diameter. In our example, this leg of the building drain will be constructed with 4-inch diameter pipe. So you must add half its diameter, 2 inches, to the depth of the trench. This calculation takes into account the amount of pipe below its centerline. The pipe will start at the foundation footing at a depth of 32 inches (30" + 2" = 32") and rise to 30½ inches (28½" + 2" = 30½") at the sweep.

Now dig the trench from the foundation wall to the sweep. Stretch a string beside the selected path to aid in keeping the trench on course. You must dig the trench at the proper slope to ensure the proper flow of wastes. Use a **spirit level** to maintain the correct grade. If you dig too deeply, the trench may require backfill. Over time, backfill may settle and cause the pipe to sag, so take care to avoid this situation. Where backfill is unavoidable, you should use brick or wet sand. Also, dig an enlarged area at the location of the sweep to provide enough space for its installation.

4.8.0 Installing the Main Stack Sweep

Locate the correct position of the sweep in the trench using a plumb bob and string suspended from the center of the sanitary tee attached to the closet bend. Note where the plumb bob comes to rest and position the sweep there. You will have to move the sweep around until the plumb bob points to the center of the sweep. Make certain that the exit of the sweep is at the proper slope and points straight down the trench. Then, brace the sweep to prevent it from moving. Pour concrete around the sweep to secure it in position. These steps ensure that the sweep will not move off center and that it will be capable of supporting the weight of the pipe above it.

Review Questions

Sections 4.0.0–4.8.0

1. The main stack for residential plumbing fixtures typically is located near the _____.
   a. tub or shower in the bathroom wall
   b. water closet in the bathroom wall
   c. lavatory in the bathroom wall
   d. kitchen sink and dishwasher

2. A properly installed stack must run _____.
   a. true vertical
   b. at less than a 5-degree slant
   c. at a slope of least ¼-inch per foot
   d. along the centerline of the building

3. You should determine the location of the stack exit through the roof by _____.
   a. measuring the distance from a continuous wall
   b. suspending a plumb bob from one floor to another
   c. suspending a plumb bob from the roof to the base of the stack
   d. measuring the length of the centerline of the stack from the basement

4. You should locate the building drain _____.
   a. perpendicular to the foundation wall and directly to the sewer main
   b. parallel to the foundation wall and directly to the sewer main
   c. at the building corner farthest from the sewer main
   d. at the building corner nearest to the sewer main

5. Measurements to locate the elevation of the main stack sweep must be made to the _____.
   a. sewer main
   b. top of the pipe
   c. bottom of the pipe
   d. centerline of the pipe

5.0.0 ◆ CHANGE OF DIRECTION OF THE BUILDING DRAIN

The building drain must change direction to connect with the floor drain and the secondary stack. Attach a sanitary wye to the outlet of the sweep for this purpose. Also, because the drain's discharge rate lessens beyond the main stack, the branch inlet of the sanitary wye will be 3 inches. The fitting is described as a 4" × 3" wye with a 4" × 3" reducer. Thus, the size of the branch of the building drain will be reduced from 4 inches to 3 inches. This adds to the economy, without affecting the efficiency of the plumbing job.

Temporarily join the sanitary wye to the exit of the sweep. Then, string a line from the secondary stack to the branch inlet of the sanitary wye. Use this line to make sure the branch inlet of the sanitary wye is pointing in the proper direction. The line also indicates the path of the trench to the secondary stack. After checking for accuracy, join the sanitary wye to the sweep.

Remember that the size of the building drain was reduced by one inch in diameter at the wye. Therefore, the trench for the next leg of the building drain must be reduced in depth. Since only half of the diameter of the pipe extends below the centerline of the pipe, the depth of the trench must be reduced by ½ inch.

5.1.0 Extending the Building Drain to the Floor Drain

With a line stretched to indicate the path of the pipe, you can trench the building drain to its point of connection with the floor drain. However, you must first calculate the rise of the drain. This will determine the elevation of the fitting that will be used to connect the floor drain.

To calculate the rise, measure the distance from the branch inlet of the sanitary wye to the point where the floor drain will connect to the building drain. This distance can be determined by consulting the plan or by measuring along the basement floor. Let's assume this distance is 21 feet.

Multiply the number of feet by the desired slope of ¼ inch per foot. You will find that the building drain will rise 5¼ inches at its connection to the floor drain (21 × ¼" = 5¼"). Using this information, dig the trench for this section of the building drain, then lay and join the pipe.

Do not fill in the trench for the building drain at this point. The piping must be visible for the plumbing inspection and not covered until the inspection is completed.

Select the fitting that will be used to connect the floor drain to the building drain. Reduce the size of the pipe from 3 inches to 2 inches above this fitting. There are two reasons for doing this. First, the building drain can be reduced at that point because it will receive the flow only from the kitchen sink. Because of the drainage capacity of the sink in fixture units, a 3-inch diameter pipe is unnecessary. Second, reducing the pipe size is more economical in terms of both material and labor. Because of its discharge rate, the floor drain must be plumbed with 3-inch pipe.

With these considerations, you should select a 3" × 2" reducer in the sanitary wye and join the fitting to the building drain. Make sure the branch inlet of the sanitary wye is properly positioned so that the floor drain branch line will drain properly.

5.2.0 Calculating the Elevation of the Secondary Stack Sweep

Before going any further, you must determine the vertical center of the secondary stack sweep. This will ensure that the building drain and the stack are properly aligned. The procedure is similar to that used to locate the sweep at the base of the main stack. Suspend the plumb bob through the center of the hole cut for the stack in the first floor soleplate. The spot at which the plumb bob comes to rest indicates the center of the stack at ground level.

With the center of the secondary stack identified, you can figure out the elevation of the sweep.

In this case, calculations can be made from the fitting used to connect the floor drain. Refer to *Figure 30*. You'll see that the stack is 8 feet away from the sanitary wye used to connect the floor drain. At a slope of ¼ inch per foot, the rise of the centerline of the building drain would be 2 inches (8 × ¼"). Remember that measurements are made to the centerline of the pipe. When the size of the pipe is reduced from 3 inches to 2 inches, the depth of the trench will be decreased by half the difference.

In our example, the difference in diameters is 1 inch. Half that distance is ½ inch. As the pipe is ½ inch smaller on the radius, the trench would begin at an elevation ½ inch less than the trench for the 3-inch line. So you must subtract ½ inch from the 2-inch rise. The building drain will rise 1½ inches from the floor drain fitting to the secondary stack sweep. Dig the trench to the proper depth and extend it to the centerline of the secondary stack, which is indicated by the suspended plumb bob.

5.3.0 Connecting the Secondary Stack Sweep to the Building Drain

After you extend the trench, you must accurately position the sweep using the plumb bob and string. With the plumb bob still suspended, place the sweep in the trench. Move the sweep around until the plumb bob points to the center of its vertical end. Pour concrete around the sweep to enable it to support the weight of the stack. With the sweep accurately located, positioned, and supported, you can install the building drain between the floor drain and the sweep.

5.4.0 Connecting the Building Drain to the Floor Drain

Refer to a plumbing plan to locate the floor drain (seasoned plumbers locate the floor drain from experience). For economy, the floor drain usually is placed near the building drain. After locating the drain, mark the path of the trench. Then, dig a trench at the proper slope and install the pipe. Join a trap to the end of the pipe to turn the drain upward to the floor drain.

The installation of the floor drain is critical, as it must be accurately positioned before the concrete floor is poured. Plumb the floor drain above the unfinished floor and support it in position. As the concrete is being poured, concrete finishers will work the concrete around the drain with a trowel. This will ensure that the floor will slope toward the drain.

5.5.0 Connecting the Building Drain to the Building Cleanout

In buildings with basements, most local codes require that a cleanout (called the *building cleanout*) be placed in the building drain just before it exits the building. The fitting may be a cleanout and test tee, sanitary tee, sanitary wye, or combination tee-wye.

Begin the installation by temporarily placing the cleanout fitting in the trench. Position the fitting so that the branch inlet of the fitting is pointing upward. You should also place the fitting an appropriate distance inside the front wall. This distance should be long enough to allow the next length of pipe to be joined to the cleanout fitting without difficulty. With the cleanout fitting located, you can then install the pipe to connect the sanitary wye at the main stack to the cleanout fitting. You must attach a vertical section of pipe to the branch inlet of the cleanout to make it accessible at surface level inside the structure.

DID YOU KNOW?
1927 – Vitreous China Fixtures

In 1927, Charles A. Lindbergh flew nonstop from New York to Paris in his airplane "The Spirit of St. Louis." In that same year, vitreous china toilets, washbasins, and drinking fountains became available, in addition to the common enameled cast-iron fixtures.

5.6.0 Connecting the Building Drain to the Sewer Main

From the cleanout, run the building sewer to the sewer main. First, establish the path of the drain with a tightly drawn string. Dig the trench with the aid of a backhoe or a trenching machine. Then, extend the building drain from the cleanout to the sewer main. Do not make the connection to the sewer main until all the plumbing in the structure is completed. This prevents sewer gases from entering the structure.

WARNING!
Be aware that toxic and flammable vapors may be present when a sewer tie-in is made. The public sewer is also a biohazard. Use the appropriate personal protective safety equipment and wear hand and eye protection as necessary.

WARNING!
The trench from the building to the sewer is usually the deepest trench you will dig on the job site. Cave-ins and falls are possible, so trench safety is extremely important. You or others may be seriously injured or even killed in a cave-in or fall.

You must use proper shoring (support the sides of the trench to prevent a cave-in). If shoring is not required, you may be required to terrace (step or shelve) the sides of the ditch. Consult the Occupational Safety and Health Administration (OSHA) requirements for restrictions on trench depth without shoring and follow OSHA guidelines on safely shoring or terracing your trench.

6.0.0 ◆ INSTALLING THE MAIN STACK

The location, height, and type of fitting that will be used to join the water closet to the stack are extremely important. The sweep at the base of the stack is accurately positioned in relation to the building drain with this fitting. The slope of fixture branch lines depends on the installed height of the fitting.

A secondary stack is required to provide a drain and vent for the kitchen sink. Use the same procedures for plumbing the secondary stack that you use for the main stack.

6.1.0 Selecting the Closet Fitting

In selecting the proper fitting, you must consider the number of fixtures and the route of their drains. From this information, you can determine the number of branches the fitting must have. Refer to the bathroom floor plan shown in *Figure 6*. You can see that it will be most efficient to connect the water closet and the bathtub to the DWV stack at the same fitting. The lavatory can later be joined to the stack at a higher elevation. Keeping the local code requirements in mind, let's say that you decide to use a sanitary tee with a side inlet. The tee is described as a 3" × 3" × 3" sanitary tee. You will use a 3" × 3" × 1½" sanitary wye to join the sanitary tee. You'll join the water closet at the end and the bathtub at the branch inlet.

6.2.0 Determining the Fitting Height

With the proper fitting selected, you can now determine its height in the DWV stack. Generally, you can determine the height of the fitting by the slope of the branch line from the water closet. The water closet drainpipe is larger than the lavatory and bathtub drain pipes, and, given an equal slope, needs more space than smaller drain lines. To determine the fall of the water closet branch line, measure the distance between the water closet and the stack and multiply that distance by the slope. In our example, that distance is 4 feet. At a slope of ¼ inch per foot, the line would fall 1 inch (4 × ¼" = 1").

Let's assume that the start of the run of pipe will begin with the pipe touching the lower surface of the floor joists. Using 3-inch pipe, this means that the centerline of the pipe would be one-half of its diameter, or 1½ inches below the floor joists. Remember, measurements are made to the pipe's centerline. This is important because the height of the fitting is determined by measurements made to the centerline of the branch inlet.

The height of the sanitary tee is in relation to the lower surface of the floor joists. The centerline of drain pipe begins 1½ inches below the floor joists and falls another 1 inch before joining the sanitary tee. Therefore, the centerline of the branch inlet of the sanitary tee must be a minimum of 2½ inches below the floor joists. This accurately locates the height of the fitting.

You will have to adjust to problems that may occur. For example, where working clearance is a problem, you may have to place the fitting at a lower height. However, you cannot allow the slope to change. To prevent this, add a short length of pipe to the vertical drain from each fixture. The assembled length of the pipe will correspond to the distance the fitting was lowered to obtain working clearance.

6.3.0 Connecting the Fitting to the Closet Flange

With the position of the sanitary tee known, you can determine the best method to couple it to the closet flange. Either a closet bend or a quarter bend and a short length of pipe can be used to connect the closet flange to the sanitary tee. Let's assume that you select a closet bend. Place the closet bend under its hole in the bathroom floor. Place blocking equal to the thickness of the finished floor under the flange. Secure the flange in position to prevent it from moving. You can use a weight or temporarily fasten the flange to the floor.

Next, insert the closet bend in the flange from below the floor level. Place the sanitary tee in its correct position. By temporarily supporting the sanitary tee and the closet bend, you can determine the correct length of the horizontal run of the closet bend. Measure the distance between the two.

If the closet bend is too long, cut it to the proper length. If it's too short, add a length of pipe between the bend and the sanitary tee. Now temporarily assemble the bend and the tee, place the assembly in its correct position, and support it from beneath.

Next, determine the correct length of the vertical run of the closet bend. If the closet bend is too long (as it will be in most cases) mark it to indicate where it should accurately join the flange and cut it to the proper length.

Sometimes, the vertical run of the closet bend will be too short, so you'll have to measure the distance between the bend and flange and cut a length of pipe to join the bend and flange.

Assemble the sanitary tee, closet bend, and the closet flange in their correct position. Then, raise the entire assembly and support it in position. Make sure the sanitary tee and the closet bend are centered in their correct positions.

6.4.0 Connecting the Sweep to the Sanitary Tee

With the sweep firmly supported in concrete and the sanitary tee temporarily supported in place, you can install the part of the stack that runs between them. As you can see in *Figure 31*, the cleanout must be positioned at floor level. Place the cleanout on the sweep, making sure that the branch of the cleanout is pointed in an accessible position. Then, join the cleanout to the sweep.

Next, connect the cleanout to the closet fitting with a length of pipe. Be sure to maintain the proper slope of the closet bend by accurately measuring and cutting this length of pipe. First, measure the distance between the stack and sanitary tee. Let's say this distance is 70 inches. Use direct measurements or manufacturer's tables to determine the distance the pipe and the sanitary tee spigot will slip down into the hub of the pipe. For a 3-inch diameter pipe, this is roughly 2½ inches per joint. You have two joints, so add the distance plus the joints (70" + 2½" + 2½"). You should cut a pipe that is 75 inches long.

A simpler method of determining this length is to hold a length of pipe next to the sanitary tee (see *Figure 32*). Place the hub next to the sanitary tee at the height it would be after the joint is made. Next, make a mark on the pipe where it would bottom in the hub of the cleanout fitting. Cut the pipe to that length. Lift the sanitary tee and closet flange off their supports to install the section of pipe. After ensuring the fit is correct, make up the joints.

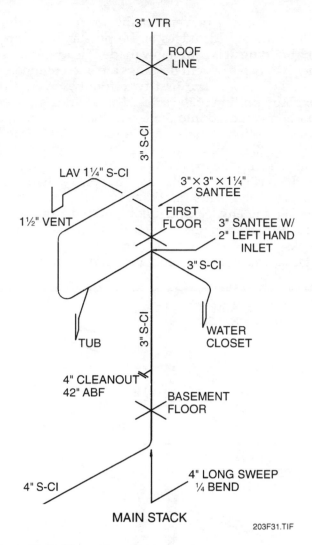

Figure 31 ◆ Cleanout position.

6.5.0 Installing the Lavatory Drainage Fitting

Before you can run the stack through the roof, you must determine the elevation of the fitting joining the lavatory branch line. This is necessary to ensure proper slope of the lavatory drain.

Begin by consulting the manufacturer's rough-in measurements for the lavatory. Refer to *Figure 3(C)*. You can see that the centerline of the drain enters the wall 18¼ inches above the floor. Remember that you must add the thickness of the finished floor to this measurement. Let's assume that the finished floor will be 1⅛ inches thick. So the centerline of the lavatory drain is 19⅜ inches off the unfinished floor (18¼" + 1⅛"). As the walls are only rough-framed at this stage of construction, mark this dimension on the nearest stud. To run the lavatory drain simply and economically, run it horizontally to the stack.

Next, determine the type of fitting you will use to connect the lavatory branch line to the

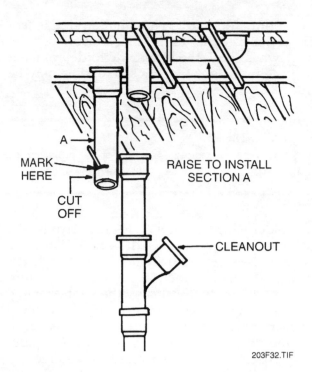

Figure 32 ◆ Determining the correct length of pipe.

stack. In this case, a 3" × 3" × 1¼" sanitary tee would be best. Now, determine the height of the fitting on the DWV stack. Start back at the lavatory drain and calculate the slope to the stack. You know that the lavatory drain is located 19⅜ inches above the surface of the roughed-in floor. From that point, the branch line will run approximately 1 foot to the DWV stack. With a slope of ¼ inch per foot, the branch line will drop only ¼ inch. This also means that the centerline of the inlet to the sanitary tee will be 19⅛ inches above the surface of the floor.

Now you can measure the length of pipe that will run from the sanitary tee at the closet bend to the sanitary tee at the lavatory drain. Use direct measurement or the manufacturer's rough-in measurements. After you cut the pipe, dry fit (trial fit) it with the sanitary tee in place to ensure that you've maintained the proper slope. If the slope is correct, join the length of pipe and the sanitary tee to the stack.

6.6.0 Extending the Stack

With all fittings used to connect branch lines in place, you can now extend the DWV stack through the roof. The procedure is simple, consisting of measuring, cutting, and joining lengths of pipe until the stack extends above the roof. The pipe should extend a minimum of six inches above the roof. Check the local code for details.

In some areas, you may reduce the diameter of the stack above the last branch line. This is mainly an economic decision. The smaller size is less expensive and easier to install, so labor costs are less. Always check the local code to determine if and to what extent the stack diameter may be reduced.

 WARNING!
Never work unprotected on a roof. Fall protection is required when working at any height over 6 feet.

In colder climates, local codes may require the addition of increasers to the stack. Sanitary increasers (see *Figure 33*) are necessary in cold climates to prevent condensing water vapor from freezing and gradually closing the vent opening. The loss of the vent could cause siphonage of the trap seals. This, in turn, would allow sewer gas to enter the structure. Refer to the local code requirements to determine if an increaser is required.

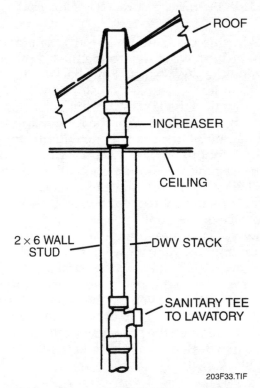

Figure 33 ◆ Sanitary increaser.

6.7.0 Flashing the Stack

The final step in running the stack is the installation of vent flashing. However, depending on the location, job, and contract, this may be a job for the plumber, the carpenter, or the roofer. Check with your supervisor, the job foreman, or the construction supervisor to see if installing vent flashing is part of your job.

Vent flashing prevents rainwater from running down the outside edges of the pipe and entering the building. It is commonly made of lead and plastic. Two types of vent flashings are in common use (see *Figure 34*). The first type totally covers the exposed portion of the stack. The flashing's upper portion is tucked into the top of the stack and the lower portion is nailed to the roof. In the second type of flashing the upper portion extends only part way up the exposed pipe. However, the hole in the flashing is smaller than the pipe's diameter, so the flashing fits tightly against the pipe, thus preventing any leaks. Caulking placed around the seal further strengthens the seal and guards against leaks. The height at which the flashing joins the stack is the main difference between the two types.

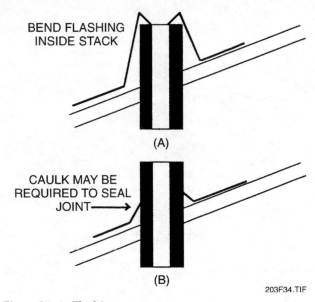

Figure 34 ◆ Flashing.

6.8.0 Connecting the Fixture Drains

After the stack has been run, you can run the various fixture drain lines. Much of this work already has been done. You've already marked the location of the fixture drains on the walls and floors. You've correctly positioned the fitting that will connect the branch line to the stack. You've calculated the slope of each branch line to determine the correct placement of the fitting. Therefore, to join the fixture drains, you only need to measure, cut, join, and brace pipe.

For the water closet, all that you need to do is to make up the joints at the stack fitting, sanitary wye, closet bend, and closet flange.

For the bathtub, measure and cut a length of drainpipe to run from the branch inlet of the sanitary wye to the bathtub drain. Then, calculate the length of pipe needed to run vertically from the horizontal drainpipe to the bathtub waste outlet.

Note the back outlet, waste, and overflow. To accurately determine this measurement, refer to the manufacturer's rough-in measurements, shown in *Figure 3(D)*. You'll find that the waste outlet of the bathtub is 1⅞ inches above the surface of the rough floor. Measure the distance from the horizontal run of pipe to an imaginary point 1⅞ inches above the rough floor level. Note that the length of both runs of pipe is measured in reference to the 90-degree ell that will join them. After cutting the pipe to the correct length, assemble the branch line starting from the sanitary tee. After final assembly, check to see that the slope is correct and that the vertical length of pipe rises 1⅞ inches off the rough floor.

Next, install the lavatory drain line. Because the wall is rough-framed, you have to mark the height of the centerline of the lavatory drain on the nearest stud. (This height is determined from the manufacturer's rough-in measurements.) When measuring the length of the horizontal run of pipe, measure to the true point of intersection with the lavatory drain. Cut a length of pipe and join it to the stack. Check the accuracy of your work.

DID YOU KNOW?

1970s – Earth-Friendly Plumbing

Caring for Mother Earth became a huge social issue in the 1970s. Grassroots and federal legislative efforts were bent on conserving our natural resources. In 1974, companies like Kohler began producing a line of low-consumption toilets, faucets, and showerheads. The products were designed to reduce the strain on municipal and private sewage systems, and lower sewage and water treatment costs. The new water closets used 3.5 gallons per flush compared with the earlier models, which required as much as 5 gallons.

By 1989, Kohler had introduced a low-flow toilet that used 1.6 gallons per flush.

Review Questions

Sections 5.0.0–6.8.0

1. Before trenching the building drain to its point of connection with the floor drain, you must first calculate the _____ of the drain.
   a. rise
   b. angle
   c. slope
   d. depth

2. In buildings with basements, most local codes require that a _____ be placed in the building drain just before it exits the building.
   a. branch line
   b. cleanout
   c. tee
   d. 90-degree ell

3. The height of the water closet fitting in the DWV stack typically is determined by the _____.
   a. elevation of the water closet
   b. heights of other fixtures in the building
   c. slope of the branch line from the water closet
   d. distance of the water closet from the building drain

4. In colder climates, local codes may require the addition of _____ to the stack.
   a. increasers
   b. reducers
   c. extensions
   d. sanitary wyes

5. _____ prevents rainwater from running down the outside edges of the pipe and entering the building.
   a. A vent cover
   b. A vent grill
   c. Vent flashing
   d. Vent channeling

DID YOU KNOW?

A Sinking Bathtub Spurs a White House Renovation

In 1948, Harry S. Truman was elected President. Under his administration, the White House underwent nearly $5 million in renovations. Several experts were in favor of tearing down the building rather than trying to correct the tangle of pipes, wires, and inefficiencies that had developed over the years. The President was spurred into action because he noticed that his bathtub was sinking into the floor.

This same year witnessed the increased use of a powder room or guest bathroom in residential building design. It was advertised as a benefit by "saving traffic through the house, making toilet training easier, and a welcome convenience for guests."

7.0.0 ◆ PLUMBING IN SLAB-ON-GRADE CONSTRUCTION

In structures built on a concrete slab, you must place the piping before the slab is poured. However, before attempting to lay out the plumbing system, consult the carpenters to determine how they will lay out the structure. Often, the north and east walls, for example, are designated as the walls from which everyone will measure. This method prevents small variations in the existing structure from affecting progress and will ensure that pipes and the walls built later are in correct relationship to one another.

Locate the point where the DWV piping will penetrate the floor from the floor plan. Do this by using a steel tape measure to lay out the measurements from a corner of the building or from a wall. Make sure that excavated dirt does not block access to areas where you need to measure. *Installation tip:* Use string to mark the front edge of plumbing walls; this enables you to stretch one tape measure parallel to the plumbing wall to measure for stacks and fixture locations. Make sure dirt is not piled where a string line might be needed.

Place **sleeves** around pipes that pass through concrete that is part of the building's structural elements, such as footings, stem walls, and grouted masonry. Sleeves, which prevent the pipe from being damaged by contact with concrete, are made from plastic pipe that is one size larger than the DWV pipe being installed. They are secured in place before the concrete is poured and must be sloped so that the pipe passing through is not pinched.

Secure the piping installation to prevent movement during backfill and compaction of the soil around the pipes. **Rebar** (a ribbed reinforcing bar) is often used for this purpose. The rebar is driven into the soil at an angle adjacent to the pipe and then tie wire is used to clamp the pipe to the rebar. Two of these bars placed perpendicular to each other provide good protection against movement.

After the plumbing under the slab is complete, the slab is poured and allowed to **cure**. Carpenters then rough-frame the structure. After the floor has cured, the walls have been erected, and the roof has been put on, plumbers are again called to the site to complete the DWV piping installation.

The installation is similar to that previously discussed for structures with basements. The major exception is that the building drain and the sweep at the base of the stack are placed before the slab is poured. Also, there is no need to bore holes in the floor as the DWV pipes already have been extended above the surface of the slab. The carpenters will bore the holes in the soleplates as they erect the walls over the piping. From the slab on up, the plumbing is a matter of locating the fixtures, running the stacks, and connecting the fixtures.

8.0.0 ◆ INSTALLING PIPE HANGERS AND SUPPORTS

You must install and support all pipe so that both the pipe and its joints remain leakproof. Improper support can cause the joints to sag, which causes added stress on the pipe and increases the probability of leaks, breaks, or even cracks between the joints. Lack of proper support can allow the drainage pipe to shift from its proper pitch and form traps. These traps fill with liquid and solid waste, causing blockage within the pipe.

8.1.0 Types of Pipe Hangers and Supports

Supports and hangers are designed to hold and support pipe in either a horizontal or vertical position. They are made from various materials including carbon steel, malleable iron, steel, cast iron, and plastic, and they come in different finishes such as copper plate, black, and electrogalvanized.

Which hanger style and finish you use depends upon the type of pipe and its application. With the greater use of plastic DWV systems in plumbing, many problems developed with the strapping and hanging of plastic pipe. In some cases, the thermal expansion and contraction of plastic DWV piping is three to five times greater than that of metal piping. The use of metal straps on DWV pipe not only abrades the surface of the plastic but also restricts the pipe from expanding and contracting. For these reasons, it is important to use hangers and supports that are properly designed for use with plastic DWV piping. *Figure 35* illustrates various plastic hangers and supports available for use with DWV piping. Since various hangers and supports are used in combination with each other, it is difficult to set up naturally exclusive categories. However, the basic components of hangers and supports can be placed into three major categories. These include pipe attachments, connectors, and structural attachments.

8.1.1 Pipe Attachments

Pipe attachments include the part of the hanger that touches or connects directly to the pipe. They may be designed for either heavy duty or light duty, for covered pipe or plain pipe.

SUSPENSION HANGER

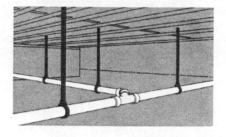

STRAP LOCK HANGER

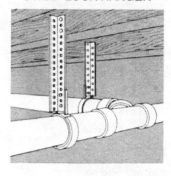

MULTI-PURPOSE HANGER

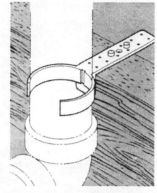

PLASTIC PLUMBERS TAPE

SNAP STRAP

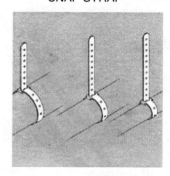

MULTI-PURPOSE HANGER

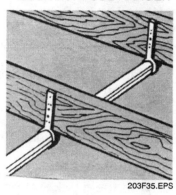

203F35.EPS

Figure 35 ◆ Plastic hangers and supports.

Several styles of hangers used for supporting piping from the ceiling horizontally are illustrated in *Figure 36*. These include various types of rings, clamps, and clevises. The styles illustrated are recommended for suspending both hot and cold noninsulated stationary piping.

Piping can be supported on wood frame construction with several styles of pipe attachments. These include pipe hooks, U-hooks, J-hooks, tube straps, tin straps, perforated band irons, half-clamps, suspension-clamps, and pipe clamps (see *Figures 37* and *38*).

CLEVIS HANGER   ADJUSTABLE SWIVEL RING HANGER   STIRRUP

203F36.TIF

Figure 36 ◆ Hangers used to support pipe from the ceiling horizontally.

U-HOOK

TIN STRAP

HOLD DOWN CLIP

203F37.EPS

Figure 37 ◆ Pipe supports for wood frame construction.

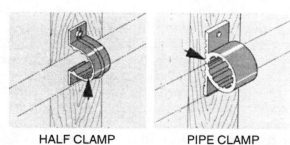

HALF CLAMP   PIPE CLAMP

203F38.EPS

Figure 38 ◆ ABS plastic pipe clamps for wood.

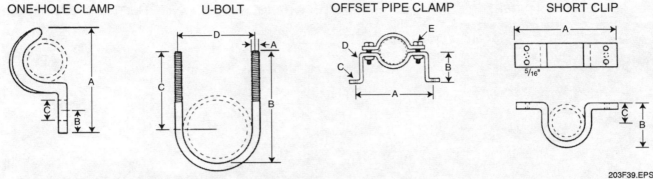

Figure 39 ◆ Attachments for use on walls or beams and columns.

Several styles of pipe attachments are available for supporting piping from the walls or the sides of beams and columns. *Figures 39* and *40* illustrate styles of pipe attachments used for this purpose. They include one-hole clamps, steel brackets, offset clamps, various styles of clips and straps, as well as combinations using extension split-clamp hangers and wall plates.

Support vertical pipe at each floor level using riser clamps (see *Figure 41*). These clamps are available in different finishes that will accommodate steel, copper, and DWV pipe. Mount the riser clamp so the bracket supports the pipe weight directly on the floor.

Other pipe attachments include universal pipe clamps (see *Figure 42*) and standard 1⅝-inch or 1 ½-inch channels (see *Figure 43*). The clamps are available in standard finishes of mild and electrogalvanized steel. Aluminum, copper-plated, and stainless steel finishes are also available from some manufacturers but usually these finishes have to be special-ordered. Insert the notched steel clamps by twisting them into position along the slotted side of the channel. The pipes can be aligned as close to one another as the couplings allow.

8.1.2 Connectors

The connector portion of the hanger is the intermediate attachment that links the pipe attachment to the structural attachment. These intermediate attachments can be divided into two groups: rods and bolts, and other rod attachments.

The rod attachments include eye sockets, extension pieces, rod couplings, reducing rod couplings, hanger adjusters, turnbuckles, clevises, and eye rods. All are available in several sizes that will meet most installation requirements.

The eye socket provides for a nonadjustable threaded connection. Use it in conjunction with a hanger rod when installing a split-ring hanger. Use extension pieces to attach hanger rods to beam clamps and other types of building attachments. They provide for a small amount of adjust-ment, approximately 1 inch, on the hanger rod (see *Figure 44*).

Rod couplings and reducing rod couplings support pipe lines where it is possible to connect to an existing stud (see *Figure 45*). You can use both to connect two pieces of threaded support rod. Use the reducing rod coupling when the support rods are of different sizes and the standard rod coupling with support rods of the same size.

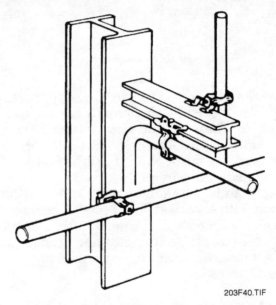

Figure 40 ◆ Clip-type attachments for beams and columns.

Figure 41 ◆ Riser clamps.

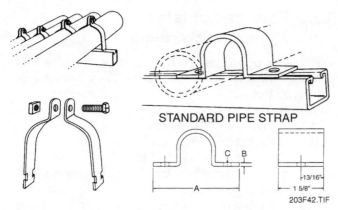

STANDARD PIPE STRAP

203F42.TIF

Figure 42 ◆ Universal pipe clamps.

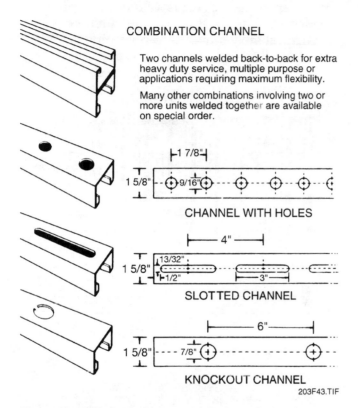

COMBINATION CHANNEL

Two channels welded back-to-back for extra heavy duty service, multiple purpose or applications requiring maximum flexibility.

Many other combinations involving two or more units welded together are available on special order.

CHANNEL WITH HOLES

SLOTTED CHANNEL

KNOCKOUT CHANNEL

203F43.TIF

Figure 43 ◆ Channels for use with universal pipe clamps.

Hanger adjusters and turnbuckles provide an adjustable threaded connection for the support rod (see *Figure 46*).

Use hanger adjusters with split-ring hangers or beam clamps when an adjustable hanger rod connection is desirable. Hanger adjusters have a permanently set swivel in the body that allows smooth adjusting during installation. Use turnbuckles to connect two hanger support rods together. They provide 6 inches of adjustment.

Use the weldless eye nut and the forged steel clevis on high-temperature piping installations (see *Figure 47*). Install the eye nut where a flexible connection is required. Use the clevis to connect the support rod to the welded lug or structural steel on heavy-duty piping installations.

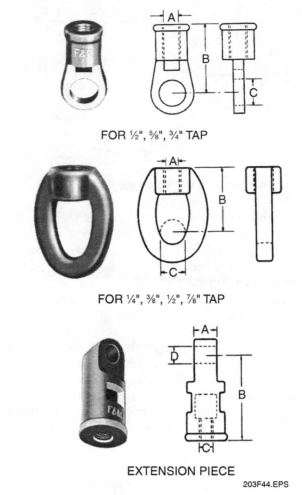

FOR ½", ⅝", ¾" TAP

FOR ¼", ⅜", ½", ⅞" TAP

EXTENSION PIECE

203F44.EPS

Figure 44 ◆ Eye sockets and extension piece.

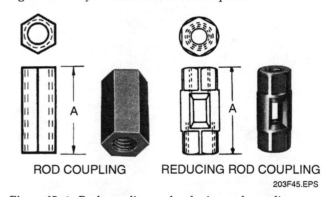

ROD COUPLING   REDUCING ROD COUPLING

203F45.EPS

Figure 45 ◆ Rod coupling and reducing rod coupling.

HANGER ADJUSTER   TURNBUCKLE

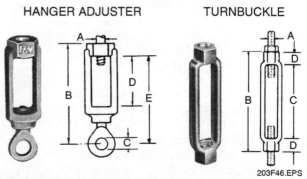

203F46.EPS

Figure 46 ◆ Hanger adjuster and turnbuckle.

3.27

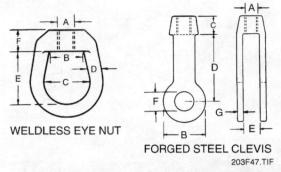

WELDLESS EYE NUT

FORGED STEEL CLEVIS

203F47.TIF

Figure 47 ◆ Weldless eye nut and forged steel clevis.

The hanger rod portion of the connector includes the eye rods and the machine-threaded rods (see *Figure 48*). Use them to connect the rod attachments to the pipe attachments to form a hanger assembly. The eye rods are available without a welded eye for lighter installations and with a welded eye for installations where more strength is required. Machine-threaded rods are available with continuous threads that run the complete length. They can be cut to the required hanger length on the job, eliminating the need for field threading. You can purchase hanger rods with two threaded ends: with right-hand threads on both ends or with one right-hand thread and one left-hand thread.

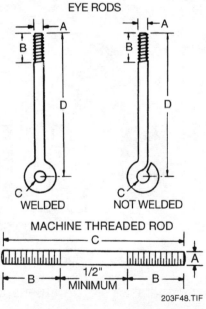

EYE RODS

WELDED   NOT WELDED

MACHINE THREADED ROD

1/2" MINIMUM

203F48.TIF

Figure 48 ◆ Hanger rods.

8.1.3 Structural Attachments

Structural attachments are the anchors and anchoring devices used to hold the pipe hanger assembly securely to the structure of the building. These include powder-actuated anchors, concrete inserts, beam clamps, C-clamps, beam attachments, brackets, ceiling flanges, plates, plate washers, and lug plates.

8.2.0 Powder-Actuated Fastening Systems

The use of powder-actuated anchor systems has increased rapidly in recent years. They are widely approved for anchoring static loads (stationary and vibration-free) to steel and concrete beams, walls, and so forth. Consult agencies such as Underwriter's Laboratories (UL) and local building codes for specific types of loads and load limits allowable in your area. In addition, OSHA requirements specify that plumbers using these systems must be licensed in their operation. Licensing can be arranged through factory sales representatives.

Powder-actuated tools drive steel pins or threaded steel studs directly into cured concrete or structural steel surfaces to hold pipe hangers, brackets, and so on (see *Figures 49* and *50*).

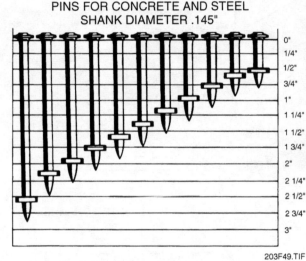

PINS FOR CONCRETE AND STEEL
SHANK DIAMETER .145"

203F49.TIF

Figure 49 ◆ Steel pins.

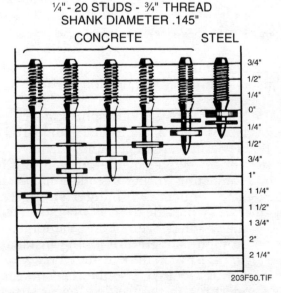

1/4" - 20 STUDS - 3/4" THREAD
SHANK DIAMETER .145"

CONCRETE   STEEL

203F50.TIF

Figure 50 ◆ Threaded steel studs.

To operate a powder-actuated fastening tool (see *Figure 51*), follow these steps:

Step 1 Feed the pin or stud into the piston.

Step 2 Feed the powder booster into position.

Step 3 Position the item to be fastened in front of the tool and press it against the mounting surface. *Note:* This pressure releases the safety lock.

Step 4 Pull the trigger handle to fire the booster charge.

The booster charge releases its pressure on the piston, and the piston forces the pin or stud into the surface and stops (see *Figure 52*).

> **WARNING!**
> Be sure to read and follow the manufacturer's instructions when using powder-actuated tools. Be sure to always wear appropriate eye protection.

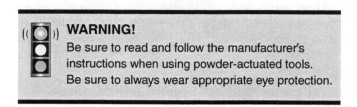

FASTENER — SPECIAL PISTON-TOOL POWDER LOAD

PISTON HAMMERS FASTENER — ENERGY TRANSFERRED TO PISTON

203F51.EPS

Figure 51 ◆ Powder-actuated tool.

Expanding gas from the safety booster drives a specially designed piston. The piston controls the speed and direction of the fastener, assuring safe penetration into the building material. Full penetration is reached when the piston is stopped inside the tool.

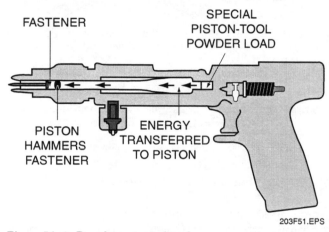

BOOSTER CHARGE

PISTON

THREADED CONCRETE STUD

BASE PLATE

CONCRETE

203F52.EPS

Figure 52 ◆ How the powder-actuated tool works.

Anchors for mounting into concrete, brick, or hollow walls are available in several styles. Use wedge anchors, sleeve anchors, stud anchors, non-drilling anchors, and self-drilling anchors to fasten the piping system to concrete (see *Figure 53*). These anchors are not recommended for use in new concrete that has not had enough time to cure.

Light-duty anchors that can be used in solid walls, hollow block, or drywall include polyset anchors, nylon anchors, plastic inserts, expansion bolts, and spring-type toggle bolts (see *Figure 54*). All are available in a variety of sizes and lengths. Always wear safety glasses when installing anchors. Be sure to use the right size drill bit and a drill that will meet the load demands of the job.

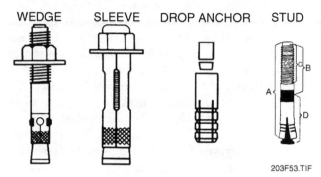

WEDGE   SLEEVE   DROP ANCHOR   STUD

203F53.TIF

Figure 53 ◆ Anchors used for concrete or brick.

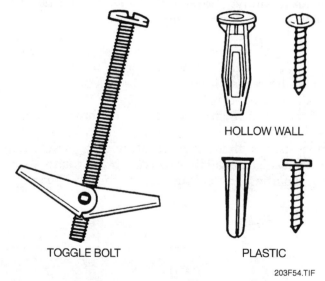

HOLLOW WALL

TOGGLE BOLT   PLASTIC

203F54.TIF

Figure 54 ◆ Light-duty anchors.

Use concrete inserts as upper attachments to suspend pipe from a concrete structure (see *Figure 55*). Standard universal steel inserts are fabricated from heavy gauge steel. They are designed with one case size that can be used with all sizes of support rods up through ¾ inches. After you install the inserts and remove the knockout plate, you can insert the special nuts. You can then suspend a hanging rod from the insert.

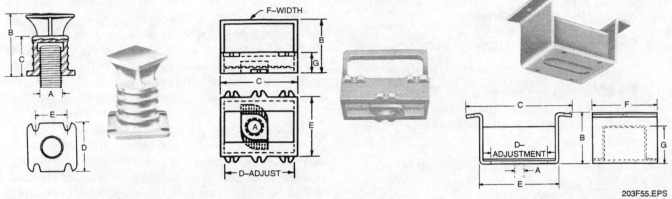

Figure 55 ◆ Concrete inserts.

Use beam clamps when the piping is to be supported from the building steel (see *Figure 56*). Attach them to the bottom flange of American Standard **I-beams.** All beam clamps are fitted with jaws that lock in position on the beam when fully and properly adjusted. Be sure to select a beam clamp that fits the thickness of the I-beam.

Use C-clamps in installations where the pipe support is to be attached to I-beams, channels, or wide flange beams when it is desirable to have the support rod offset from the beam. Secure the clamp to the flange with a cup-pointed, hardened setscrew. After the clamp is in place, thread a support rod into the tapped hole at the base. You can install retaining straps with the clamp to prevent the C-clamp from moving.

Use beam attachments to attach support rods to a structural member. You can weld or bolt them in either an upright or inverted position. When you install them in an inverted position, you can make small vertical adjustments using the hanger rod.

Brackets and clips used to attach threaded connections to wood beams are available in several styles. Secure them in place using bolts or screws. You may also install them on concrete structures by either bolting or welding them to the steel beam.

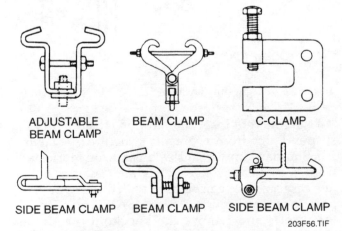

Figure 56 ◆ Beam clamps for supporting piping from the building steel.

Ceiling flanges and plates are recommended for suspending pipelines from wood beams or ceilings (see *Figure 57*). The ceiling plate gives a finished appearance where the rods and pipe enter the ceiling.

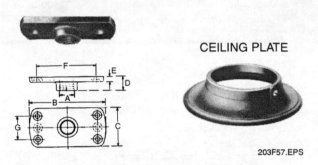

Figure 57 ◆ Ceiling flange and plate for suspending pipelines from wood beams or ceilings.

Use structural attachments such as washer plates, lug plates, or clevis plates to suspend support rods (see *Figure 58*). The heavy-duty washer plate is used on top of channels or angles to support the pipe with rods or U-bolts. When the support rods are to be suspended from concrete ceilings, use lug plates.

When using structural attachments, you should do the following:

- Check the specs (specifications) to see what type of hanger is required.
- Check pre/post-stressed concrete T-sections to make sure that installing the anchors will not weaken the structure.
- Clear all locations for concrete inserts with the architect or structural engineer in charge of the building.
- Check the specs before welding attachments onto the structure.
- Avoid using powder anchors.
- Make sure that the anchor is properly inserted.

CONCRETE CLEVIS PLATE   STEEL WASHER PLATE   CONCRETE SINGLE LUG PLATE

203F58.EPS

Figure 58 ◆ Structural attachments for suspending support rods.

8.3.0 Special Hangers and Accessories

Use pipe rollers to support piping that is subject to expansion or sideways movement because of temperature changes (see *Figure 59*). These rollers allow the pipes to expand and to move laterally. You can also use rollers as a rolling guide to feed the piping into place during pipe system assembly.

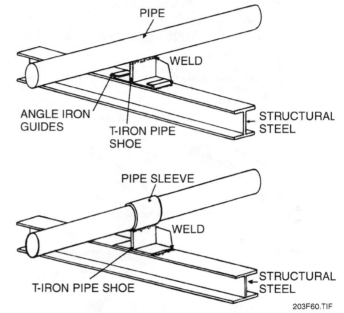

Figure 60 ◆ Field-made pipe alignment guides.

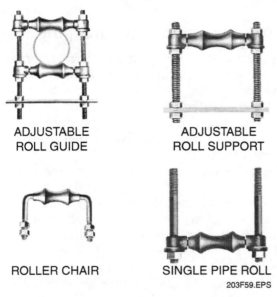

ADJUSTABLE ROLL GUIDE   ADJUSTABLE ROLL SUPPORT

ROLLER CHAIR   SINGLE PIPE ROLL

203F59.EPS

Figure 59 ◆ Pipe rollers.

Use field-made pipe alignment or shoe guides, which allow the pipe to expand and contract, to assist in pipe assembly. Two types of shoe guides you can make from materials readily available on the job site are shown in *Figure 60*. The three parts of this system are the pipe shoe, shoe guide, and pipe sleeve.

Install spring hangers at hanger points where vertical thermal movement occurs (see *Figure 61*). They are available in several styles and weights. Light-duty spring hangers are designed to provide a flexible spring support for light-duty loads where the vertical movement does not exceed 1¼ inches. In installations where you want to prevent vibration noise and other sounds from being

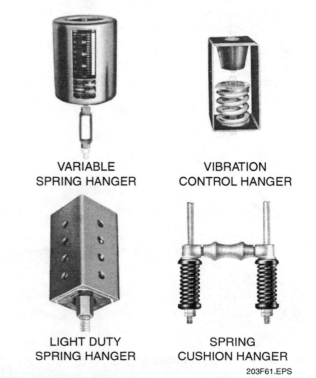

VARIABLE SPRING HANGER   VIBRATION CONTROL HANGER

LIGHT DUTY SPRING HANGER   SPRING CUSHION HANGER

203F61.EPS

Figure 61 ◆ Spring hangers.

transmitted into the building, use vibration control hangers to suspend the piping.

Spring cushion hangers are designed for use with a single pipe run in installations where the vertical movement does not exceed 1¼ inches. Use constant support hangers where constant, accurate support is needed on piping systems that move vertically because of temperature changes. It is a good practice to first install the pipe with rigid hangers. After the pipe is installed, replace the rigid hangers with the proper spring hangers.

Devices that protect the pipe in the hangers include protection saddles and insulation protection shields (see *Figure 62*). Protection saddles are designed for use on high-temperature lines or where heat losses are to be kept at a minimum. They are also designed to transmit the pipe load to the supporting unit without damaging the covering. Insulation protection shields are recommended when low compression strength and vapor-barrier-type insulation such as foam or fiberglass is installed. The protection shield prevents the hanger unit from cutting, crushing, or otherwise damaging the insulation or the vapor barrier.

8.4.0 Pipe Hanger Locations

Code governs pipe hanger locations. Factors governing pipe hanger locations are pipe size, piping layout, concentrated loads of heavy valves and fittings, and structural building steel available for piping support.

Support concentrated piping loads as close as possible to the load. Support terminal points close to the equipment. Locate hangers right next to any change of piping direction.

When installing pipe supports and hangers, you should be familiar with the engineer's specifications as well as local plumbing codes. For sprinkler system installation, you must be familiar with the specifications set up by the National Fire Protection Association (NFPA).

8.5.0 Supporting Vertical Piping

Support vertical piping at sufficient intervals to keep the pipe in alignment. Methods for supporting vertical piping depend upon whether the pipe is to be supported at or between the floor line as well as the size of the pipe being supported. Vertical support systems will also depend upon local code requirements.

You can support vertical pipe at each floor line with riser clamps (refer to *Figure 41*). It may also be necessary to install hangers that are attached between floors to the walls or vertical structural members. These hangers will maintain alignment and will also support part of the vertical load of the pipe. These vertical runs between floor levels can be supported using either an extension ring hanger and wall plate (see *Figure 63*) or a one-hole strap (see *Figure 64*).

8.6.0 Supporting Horizontal Piping

You must support horizontal or sloping pipe at intervals that are close enough to prevent sagging and to keep the pipe in alignment. Sagging pipes produce traps that allow deposits to accumulate in low spots. Traps also can allow air or vapor to accumulate in high spots. As a general rule, each length of pipe should be independently supported so that it does not have to depend on the neighboring pipe for support. Calculate the slope on which supports are placed and the distance between supports so that each point of support is lower than the nearest upstream point (see *Figure 65*).

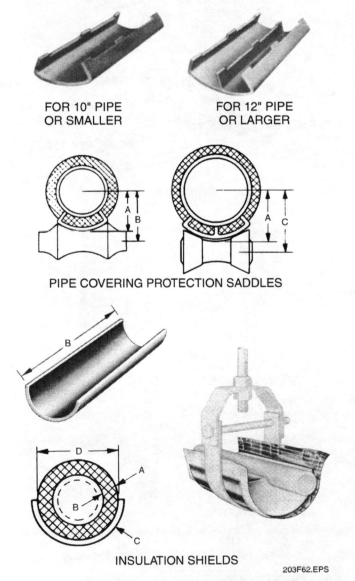

FOR 10" PIPE OR SMALLER

FOR 12" PIPE OR LARGER

PIPE COVERING PROTECTION SADDLES

INSULATION SHIELDS

203F62.EPS

Figure 62 ◆ Protection saddles and insulation shields.

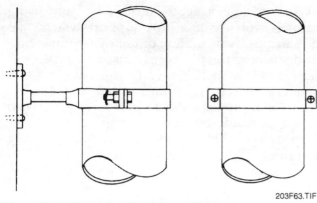

Figure 63 ◆ Extension ring hanger.

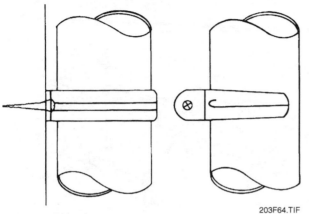

Figure 64 ◆ One-hole strap.

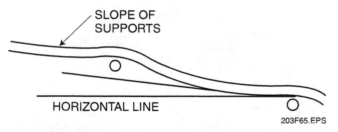

Figure 65 ◆ Slope supports.

8.7.0 Supporting Closet Bends

You must support closet bends horizontally and vertically to prevent movement in either direction. You can use a clevis hanger (see *Figure 66*) or a wood frame brace (see *Figure 67*). The type of clevis or strapping material used will depend upon the type of pipe being supported.

8.8.0 Supporting Stack Bases

You must provide adequate support at the stack base. When the stack base is underground, you can support it by placing a brick or concrete support under the fitting at the base of the stack (see *Figure 68*). Support aboveground stack bases with a hanger placed on the base fitting or as close to it as possible (see *Figure 69*).

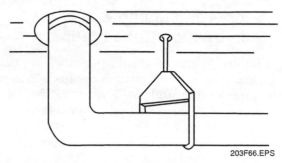

Figure 66 ◆ Closet bend supported using a clevis hanger.

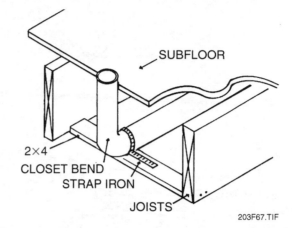

Figure 67 ◆ Closet bend braced in wood frame construction.

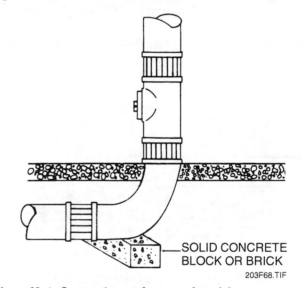

Figure 68 ◆ Supporting underground stack bases.

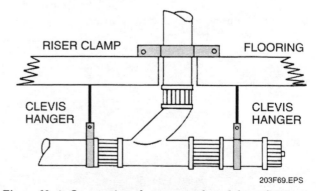

Figure 69 ◆ Supporting aboveground stack base fittings.

8.9.0 Supporting Multiple Side-by-Side Runs of Pipe

In installations where multiple pipelines are run side-by-side, you can use various methods to support the pipe in one unit. Several ways that you can support multiple horizontal or vertical runs of pipe using either a frame or trapeze assembly are shown in *Figures 70, 71, 72,* and *73.* These assemblies make it possible to hang, attach, mount, frame, and support the piping system in one unit.

When you construct the trapeze or frame using channels, you create a pipe support system that you can modify by adding or removing pipe without disturbing the previously installed pipe. Several ways in which you can use strut clips and channels when hanging and supporting side-by-side runs of pipe are shown in *Figure 74.* You can also support side-by-side runs of pipe directly on a trapeze assembly not made from channels, using hold-down clips (see *Figure 75*).

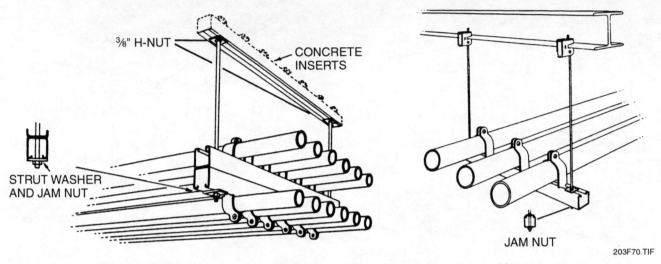

Figure 70 ◆ Typical applications using trapeze hangers made from one or more channel assemblies.

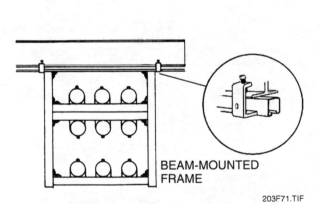

Figure 71 ◆ Beam-mounted channel frame using beam clamp.

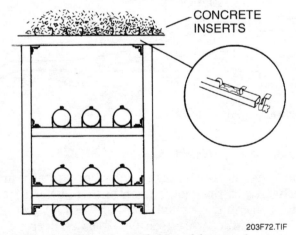

Figure 72 ◆ Ceiling-mounted channel frame using concrete inserts.

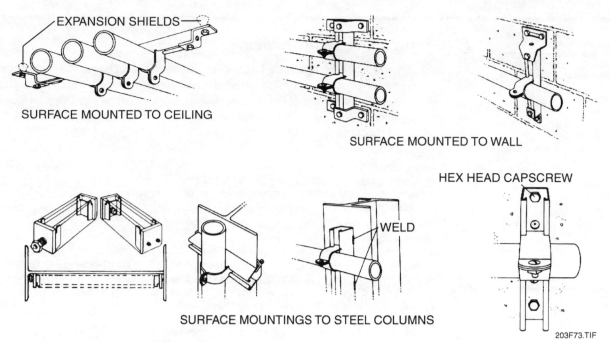

SURFACE MOUNTED TO CEILING

SURFACE MOUNTED TO WALL

HEX HEAD CAPSCREW

WELD

SURFACE MOUNTINGS TO STEEL COLUMNS

203F73.TIF

Figure 73 ◆ Surface-mounted channel frames.

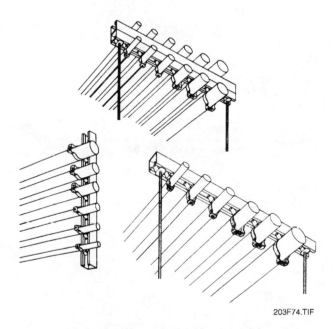

203F74.TIF

Figure 74 ◆ Side-by-side runs of piping supported with strut clips and channels.

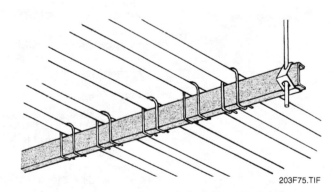

203F75.TIF

Figure 75 ◆ Supporting piping directly on steel or wood trapeze using multipurpose clips.

Plumbing Code Requirements

Plumbing codes regulate the maximum interval allowed between supports. Plumbing codes vary from area to area, but each may be based on one of several model codes. Check your local code. Following is a sample of model code requirements for hanger and support locations as they apply to cast-iron pipe, threaded pipe, copper tubing, lead pipe, and plastic pipe.

The Uniform Plumbing Code (UPC)

Vertical Piping
- Cast-iron pipe: Every story or closer
- Threaded pipe: Not less than every other story
- Copper tubing: Each story or at maximum intervals of 10 feet
- Lead pipe: Intervals not exceeding 4 feet
- Plastic pipe: Not addressed

Horizontal Piping
- Cast-iron pipe: Not more than 5-foot intervals where joints occur
- Pipe exceeding 5 feet in length may be supported at intervals not exceeding 10 feet
- Supports must be within 18 inches of the hub or joint
- For hubless or compression-gasketed joints, support must be at least every other joint unless the length between supports exceeds 4 feet—then support must be provided at each joint.
- Each horizontal branch connection must be supported
- Threaded pipe: 10-foot intervals for ¾ inch and smaller; 12-foot intervals for 1 inch and larger
- Copper tubing: 6-foot intervals for 1½ inches and smaller; 10-foot intervals for 2 inches and larger
- Lead pipe: Continuous support along entire length
- Plastic pipe: Not to exceed 4 feet

The Standard Plumbing Code (SPC)

Vertical Piping
- Cast-iron pipe: Every story level and at intervals not to exceed 15 feet
- Threaded pipe: Not less than every other story at intervals not to exceed 30 feet
- Copper tubing: Each story for 1½ inches and over; not more than 4-foot intervals for 1¼ inches and smaller

Horizontal Piping
- Cast-iron pipe: 5-foot intervals on 5-foot lengths; 10-foot intervals on 10-foot lengths
- Threaded pipe: Not to exceed 12 feet
- Copper tubing: 6 feet for 1½ inches and smaller; 10 feet for 2 inches and larger
- Lead pipe: Continuous support along entire length
- Plastic pipe: Not to exceed 4 feet

National Plumbing Code

Vertical Piping
- Cast-iron pipe: Not to exceed 15 feet
- Threaded pipe: Not to exceed 15 feet
- Copper tubing: Not to exceed 10 feet
- Lead pipe: Not to exceed 4 feet
- Plastic pipe: Not to exceed 4 feet

Horizontal Piping
- Cast-iron pipe: Not to exceed 5 feet
- Threaded pipe: Not to exceed 12 feet
- Copper tubing: 6 feet for 1¼ inches and smaller; 10 feet for 1½ inches and larger
- Lead pipe: Continuous support along entire length
- Plastic pipe: Not to exceed 4 feet

Other Requirements

The Manufacturers Standardization Society of the Valve and Fitting Industry publishes a standard that is often referred to by the model building codes when it comes to hangers and supports. This standard, which is the industry source, is titled *Pipe Hangers and Supports: Selection and Application* and is available from the Society as SP-69-83.

In addition, some of the model plumbing codes also contain information about specialty piping. You must know and follow the governing plumbing code.

9.0.0 ◆ GRADE

DWV piping systems rely on gravity to move solid and liquid wastes, so they must be installed at a slope toward the point of disposal. In the plumbing industry, this slope is called *grade*, and DWV systems are designed with the grade engineered into the system. Architects and engineers who design a piping system determine the grade. However, in some cases, such as in residential plumbing, the plumber specifies the grade.

9.1.0 The Importance of Grade

In a system with the proper grade, the velocity, or speed, of the flowing liquid wastes will scour the insides of the pipe, carrying the solids away. However, if the grade is too steep, the liquid wastes may flow too fast, leaving the solids behind. If the grade is too shallow, the liquid wastes will not flow fast enough to scour the pipe and remove the solid wastes. In either case, solid wastes will soon obstruct the pipe.

If the grade of a line of pipe is unnecessarily changed, blocked pipes will result. For this reason, proper grade, once established, must be held constant.

9.2.0 Grade Information

You can obtain information about the proper grade for the job from local plumbing codes, job specifications, or directly from plans or blueprints. Local plumbing codes specify the required grade for various pipe sizes and plumbing applications. Codes may also require the grade to be held constant throughout the plumbing job. If the grade is not specified on prints, specs, or by the local code, contact the local plumbing inspector for this information.

9.2.1 Velocity Tables

Local codes may require that the flow through a piping system maintain a minimum velocity, usually 2 feet per second (fps). Usually, the code includes a velocity table (see *Figure 76*) that you can use to determine the approximate velocity of flow through a pipe system of a given size installed with a given fall. Notice that a 2-inch pipe requires a fall of ¼ inch per foot to obtain a velocity of 2 fps while an 8-inch pipe requires a fall of only ¹⁄₁₆ inch per foot to obtain the same velocity.

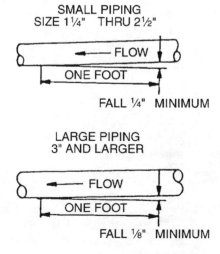

APPROXIMATE VELOCITY IN FT./SEC. AT INDICATED FALL/FT.

Pipe Size	¹⁄₁₆ in./ft.	⅛ in./ft.	¼ in./ft.	½ in./ft.
2 in.	1.02	1.44	2.03	2.88
3 in.	1.24	1.76	2.49	3.53
4 in.	1.44	2.03	2.88	4.07
5 in.	1.61	2.28	3.53	4.56
6 in.	1.76	2.49	4.07	5.00
8 in.	2.03	2.88	4.23	5.75
10 in.	2.28	3.23	4.56	6.44

203F76.EPS

Figure 76 ◆ Pipe slopes and flow velocity.

9.2.2 Specifications

Grade information is provided in the specifications, or specs, for a structure. Be aware that job specs contain information on many aspects of the job, including plumbing, structural, electrical, and heating and ventilation (see *Table 1*).

To determine the grade from the specs, you must look under *plumbing,* then look under *storm water* or *sanitary sewer system* (whichever is applicable) and then look for *grades and elevations.* This section will specify the grades and elevations you must use for the job, along with other important information.

9.2.3 Blueprints

You can also find grade information in architectural blueprints. Plot plans, which are included in the set of prints, show the location of the structure on the lot as well as dominant features and the elevation at various points on the lot (see *Figure 77*). On the plot plan you can find the finished elevation of the sewer manhole cover, the **invert elevation**, and the slope of the surrounding terrain.

Table 1 Specifications Sample

> 10. Sanitary Sewer Systems
>   A. Description: . . .
>   B. Materials and Fittings: . . .
>   C. Grades and Elevations: Sewer lines shall be uniformly graded to the elevations shown. If no elevations are given, sewers shall be pitched not less than ⅛ inch per foot.
>   D. Fixture connections: . . .

9.2.4 Elevations

To determine the elevation, locate a **bench mark**, or reference point. Bench marks are permanent markers set at specific elevations. You can mark the elevation directly on top of the bench mark. You and others on the job site will use the bench mark as a reference. In situations where a bench mark cannot be located, some other point of known elevation must be located. That point is often located from blueprints.

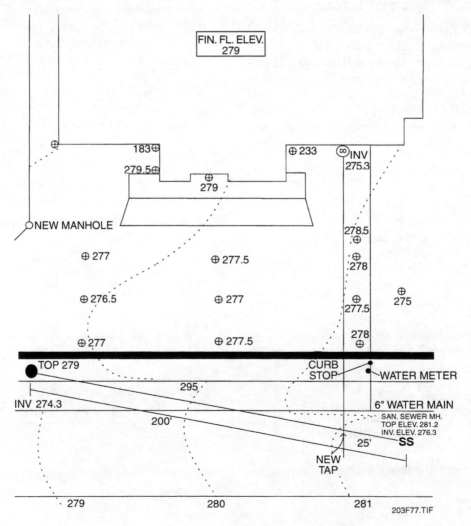

Figure 77 ◆ Plot plan.

To make reading and referring to local elevations easier, the county engineer or surveyor establishes local bench marks, which are usually set at 0' or 100'. In addition, an architect often specifies a print bench mark. The print bench mark is referenced to either a U.S. or local bench mark and makes reading that set of prints simpler.

9.3.0 Calculating Grade

Grade is the slope or fall of a line of pipe in reference to a horizontal plane. Plumbers use grades from ⅛ inch per foot to ½ inch per foot. Designers specify grades such as ¹⁄₁₆ inch per foot and ¹⁄₃₂ inch per foot for large pipelines and other special applications. Grade is expressed as vertical fall in fractions of an inch per foot of horizontal run. For example, a grade of ¼ inch means that the line of pipe will fall ¼ inch for every foot of run. The *fall* is the total change in elevation for the length of pipe. The *run* is the horizontal length of the run of pipe.

When you work with fall (F), grade (G), and run (R), remember the following:

- Fall is expressed in inches.
- Grade is expressed in inches per foot.
- Run is expressed in feet.

Depending upon which of the components you need to calculate, use one of the following formulas:

- Fall = Grade × Run (F = G × R)
- Grade = Fall ÷ Run (G = $\frac{F}{R}$)
- Run = Fall ÷ Grade (R = $\frac{F}{G}$)

To see how grade is determined, look at *Figure 78*. The horizontal run of the line of pipe is 40 feet. Its fall is 10 inches. Therefore, the grade is ¼ inch per foot.

G = $\frac{F}{R}$

G = $^{10"}/_{40'}$ = .250 or ¼" per foot

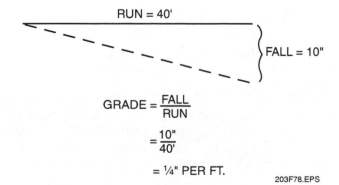

GRADE = $\frac{FALL}{RUN}$

= $\frac{10"}{40'}$

= ¼" PER FT.

203F78.EPS

Figure 78 ◆ Figuring grade.

Note that the answer first appears in decimal form. You must convert this to a fraction. You could use a conversion table (see *Table 2*). However, your job will be easier if you memorize the common fractional decimal equivalents. They are as follows:

¹⁄₁₆ = .0625
⅛ = .125
¼ = .250
⅜ = .375
½ = .500

Test your knowledge:

If R = 28' and F = 7", what is G?

The answer is .25 or ¼ inch per foot.

Table 2 Decimal Equivalents of Standard Fractions

1/64	.015625	33/64	.515625
1/32	.03125	17/32	.53125
3/64	.046875	35/64	.546875
1/16	.0625	9/16	.5625
5/64	.078125	37/64	.578125
3/32	.09375	19/32	.59375
7/64	.109375	39/64	.609375
1/8	.125	5/8	.625
9/64	.140625	41/64	.640625
5/32	.15625	21/32	.65625
11/64	.171875	43/64	.671875
3/16	.1875	11/16	.6875
13/64	.203125	45/64	.703125
7/32	.21875	23/32	.71875
15/64	.234375	47/64	.734375
1/4	.25	3/4	.75
17/64	.265625	49/64	.765625
9/32	.28125	25/32	.78125
19/64	.296875	51/64	.796875
5/16	.3125	13/16	.8125
21/64	.328125	53/64	.828125
11/32	.34375	27/32	.84375
23/64	.359375	55/64	.859375
3/8	.375	7/8	.875
25/64	.390625	57/64	.890625
13/32	.40625	29/32	.90625
27/64	.421875	59/64	.921875
7/16	.4375	15/16	.9375
29/64	.453125	61/64	.953125
15/32	.46875	31/32	.96875
31/64	.484375	63/64	.984375
1/2	.5	1	1.0

Use grade to ensure that you dig the trench for the pipeline at the correct depth. First, calculate the fall. Fall is equal to grade multiplied by run (see *Figure 79*). The line of pipe runs a horizontal distance of 16 feet at a grade of ¼ inch per foot. Therefore, the fall is 4 inches.

$$F = G \times R$$
$$F = \text{¼" per foot} \times 16' = 4"$$

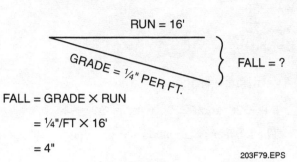

FALL = GRADE × RUN

$$= \text{¼"/FT} \times 16'$$
$$= 4"$$

203F79.EPS

Figure 79 ◆ Figuring fall.

Test your knowledge:

If G = ¼" per foot and R = 50', what is F?

The answer is 12.5".

Once you know both grade and fall, you can determine the horizontal run of the pipe without actually measuring. Run equals fall divided by grade (see *Figure 80*). The fall in this figure is 8 inches. The grade is ¼ inch per foot. Therefore, the run is 32 feet.

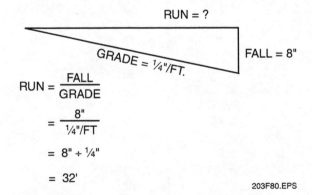

$$RUN = \frac{FALL}{GRADE}$$
$$= \frac{8"}{\text{¼"/FT}}$$
$$= 8" \div \text{¼"}$$
$$= 32'$$

203F80.EPS

Figure 80 ◆ Figuring run.

$$R = \frac{F}{G}$$
$$R = 8" \div \text{¼" per foot} = 32'$$

Note that you can convert the fraction ¼ to a decimal (0.250) to do this calculation.

Test your knowledge:

If G = ⅛" per foot and F = 13", what is R?

The answer is 104'.

9.4.0 Calculating Percent of Grade

Percent of grade is a second way to describe the slope of a line of pipe. It is used primarily for sewer and water mains and for large diameter pipelines. The procedure used to calculate the percent of grade is similar to that used for calculating grade. Percent of grade is equal to the fall in feet, divided by the run in feet, multiplied by 100. Multiplying by 100 transforms $\frac{F}{R}$ into a percentage figure. When working with greater distances, you'll find that it's much easier to work with percent of grade than with grade. To see how to calculate percent of grade (PG), look at *Figure 81*. The fall is 10 feet. The run is 1,000 feet. Therefore, the percent of grade is 1%.

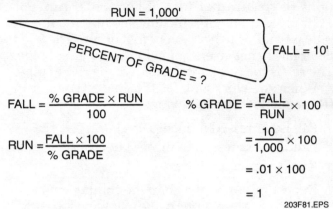

$$FALL = \frac{\text{\% GRADE} \times RUN}{100}$$

$$RUN = \frac{FALL \times 100}{\text{\% GRADE}}$$

$$\text{\% GRADE} = \frac{FALL}{RUN} \times 100$$
$$= \frac{10}{1,000} \times 100$$
$$= .01 \times 100$$
$$= 1$$

203F81.EPS

Figure 81 ◆ Figuring percent of grade.

$$PG = \frac{F}{R} \times 100$$
$$PG = 10'/1,000' = .01 \times 100 = 1\%$$

Refer to *Figure 81*. Fall and run can also be calculated with derived formulas, as shown.

Test your knowledge:

If PG = 2% and R = 500', what is F?
The answer is 10'.

If PG = 3% and F = 30", what is R?
The answer is 83.33'.

To get this answer, you must first convert F to feet (30" = 2.5'), then divide that result by PG.

9.5.0 Measuring Grade

You must install drainage and waste piping systems at a specified, constant grade. To ensure accuracy in laying out and measuring grade, you will use one or more of the following leveling tools:

- General-purpose level
- Torpedo level
- Line level
- Plumber's level

- Builder's level and transit level
- Cold beam laser

The length of pipe to be laid and the complexity of the job determine which tool should be used.

Leveling tools rely on a **spirit vial**, a small, transparent container that is partially filled with a fluid, or spirit. A small air space is left in the vial. When the vial is placed horizontally, the air space appears as a bubble. If the vial is placed on a perfectly level surface, the bubble is centered. If the vial is placed on an inclined surface, the bubble will move toward the higher end of the vial. Usually, two lines are marked across the vial. This adds to the accuracy of the device by providing reference points.

Many different leveling tools incorporate the spirit level. The most common, especially among plumbers and carpenters, is the general-purpose level (see *Figure 82*). Levels used in plumbing should be made with an aluminum-magnesium alloy so they won't warp from moisture—a common problem with wood-bodied levels. This type of level has three vials. The two end vials measure vertical alignment, such as DWV stacks. The center vial is used to level horizontally, as in a run of pipe.

Figure 82 ◆ General-purpose level.

Laying pipe to a specific grade with the general-purpose level has one major drawback. The bubble will move toward the higher end of the level, but will not indicate the precise grade. Using this level, a plumber can say only that the grade is one-fourth of a bubble or one-half of a bubble. For this reason, you should use more accurate devices wherever possible.

If you must use the general-purpose level, you can modify it to make it more suitable for measuring grade. First, determine the grade to be used in the system. For example, assume a grade of ¼ inch per foot. Also, assume that your level is 36 inches (3 feet) long.

You can multiply the grade by the length of the level (¼" × 3') to determine that the line of the pipe will fall ¾ of an inch over the length of the level. You can use this figure to simulate the fall of the line of pipe. To do this, place a ¾-inch wood block under one end of the level (see *Figure 83*).

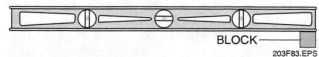

BLOCK

203F83.EPS

Figure 83 ◆ Modified level.

Now observe the center vial of the level to determine how much of a bubble (⅛, ¼, ½) crosses the indicator mark. You can then lay the pipe, making sure the same amount of bubble is held with each additional length.

A simpler method is to tape the wood block to the bottom of the level. Note that the piece of wood must be located precisely at the end. Otherwise, the grade will not be accurate. To use this device, place the modified level on the length of pipe (see *Figure 84*). Place the modified end on the downward end of the pipe. If the trench is dug at the proper grade, the bubble should center itself between the two indicator marks. In effect, the level will indicate that the line of pipe is level because the wood block compensates for the fall of the line of pipe.

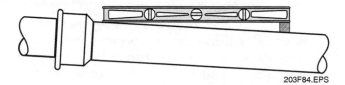

Figure 84 ◆ Using the modified level.

The torpedo level is similar to the general-purpose level (see *Figure 85*). However, it is much smaller and much more streamlined. The major advantage of this tool is in measuring grade for small runs of pipe, such as fixture branch lines that run a short distance to the stack. The torpedo level is also light and easy to manipulate in tight places.

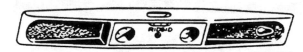

Figure 85 ◆ Torpedo level.

The line level is a small **spirit level** (see *Figure 86*). Hooks on both ends allow you to hang it on a tightly drawn string line. When using the line level, first locate and place stakes beside the desired path of the pipe. Tie a length of string to the first stake, stretch it, and secure it to the second stake. Then hook the line level onto the string. Read the portion of the bubble over an indicating mark to determine the grade. Raise or lower the

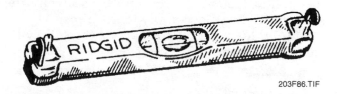

Figure 86 ◆ Line level.

string at one end until you achieve the desired grade. With the desired grade indicated by the tightly drawn string, install the line of pipe at the specified grade.

The plumber's level (*Figure 87*) is similar to the general-purpose level. Its primary advantage is that one of the spirit vials can be rotated. After determining the grade to be used, you can rotate the vial to the desired slope and lock the vial. When the level is placed on the line of pipe, the bubble should be centered within the indicating marks on the vial.

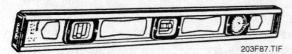

Figure 87 ◆ Plumber's level.

A second style of plumber's level includes a vial mounted within a movable pointer (see *Figure 88*). The corresponding scale is marked off in percent of grade. If you want a grade of ¼ inch per foot, loosen the setscrew and move the indicator to the fourth ⅟₁₆-inch mark. Then, tighten the setscrew and place the level on a line of pipe. When the line of pipe is at the proper grade, the bubble will be centered within the vial on the pointer.

Figure 88 ◆ Plumber's level with a movable pointer.

Plumbers use the builder's level (*Figure 89A*) when laying pipe over great distances. Sometimes referred to as a *dumpy level*, it is used to measure elevations and horizontal angles. The transit, *Figure 89(B)* can be tilted upward to check for plumb on vertical walls or vertical runs of pipe and to measure angles from vertical. The transit allows you to perform more functions, so it is more expensive than the builder's level. Plumbers generally use the builder's level, because they are mainly concerned with elevations and grades. Because of this, the following discussion focuses on the builder's level. (The same procedures also apply to the transit.)

The builder's level consists of the following seven basic parts:

- Telescope
- Leveling vial
- Leveling screws
- Protractor circle
- Index
- Tripod
- Plumb bob

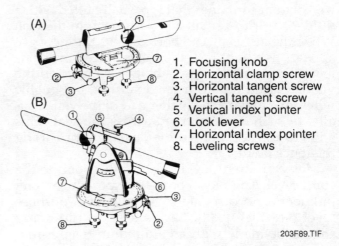

1. Focusing knob
2. Horizontal clamp screw
3. Horizontal tangent screw
4. Vertical tangent screw
5. Vertical index pointer
6. Lock lever
7. Horizontal index pointer
8. Leveling screws

Figure 89 ◆ (A) Builder's level. (B) Transit level.

The telescope is the fundamental part of the builder's level. You can measure elevations and angles through the line of sight established by the telescope. The line of sight is weightless and continuous; unlike a string line, it won't sag and is therefore more accurate (see *Figure 90*).

The telescope is designed to magnify the image so that you can more clearly see the object, or target. A magnification of 20× (power) is common, with scopes of larger power available. As a general rule, when the job involves greater working distances, use a greater magnification power.

PERFECTLY STRAIGHT AT ALL POINTS

Figure 90 ◆ Line of sight.

Cross hairs increase the telescope's accuracy by providing reference marks (see *Figure 91*). The horizontal cross hair is used in reading elevations. The vertical cross hair is used to check the plumb of objects.

A leveling vial is attached directly to the telescope, which must be leveled in reference to the leveling vial to ensure accuracy. The vial is designed like a spirit level, but is much more accurate. Use the leveling screws to center the bubble in the leveling vial as shown in *Figure 92*. These also ensure the accuracy of the telescope. (Use of the leveling screws is discussed in detail later in this section.)

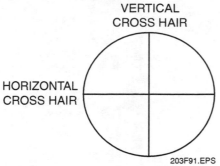

Figure 91 ◆ Cross hairs.

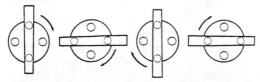

FOR A FINAL LEVEL CHECK, rotate the telescope over each of four leveling points to be sure the bubble remains centered.

The Golden Rule for quick and simple leveling is THUMBS IN, THUMBS OUT. Turn BOTH screws equally and simultaneously. Practice will help you get the feel of the screws and the movement of the bubble. It will also help to remember that the direction your left thumb moves is the direction the bubble will move.

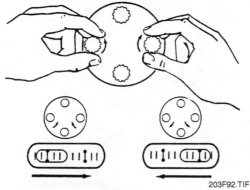

Figure 92 ◆ Leveling screws.

Use the protractor circle to accurately adjust the telescope and to measure horizontal angles (see *Figure 93*). The circle is divided into 360 degrees. The telescope is positioned so that it can rotate above the circle. An indicator accurately determines or measures horizontal angles. A **vernier** indicator frequently is used to increase accuracy.

The index, which is a pointer, is attached to the telescope. It includes the protractor circle and an indicator. The base is stationary and marked off in degrees from 0 to 360. The indicator rotates over these markings from a center pivot. This feature makes it easy to take 90-degree and 45-degree views and measurements.

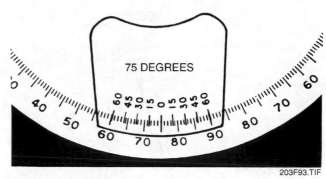

Figure 93 ◆ Protractor circle.

The builder's level is mounted on top of a tripod (see *Figure 94*). You can adjust each of the tripod's three legs individually so that you can set up the level on soft or irregular ground.

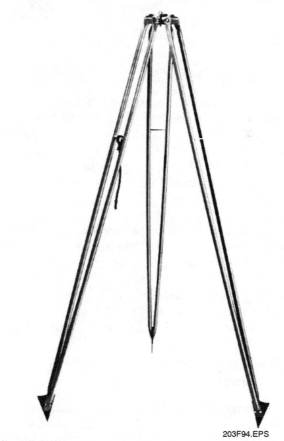

Figure 94 ◆ Tripod.

The plumb bob is supported from a special hanger at the base of the tripod (see *Figure 95*). It is used when the builder's level must be set up over a specific point, such as a monument, or a stake at the corner of a lot. To center the builder's level over the desired point, move the legs of the tripod around until the plumb bob is over the desired point. Some builder's levels have a fine adjusting screw that may be used to accurately center the instrument over the desired point.

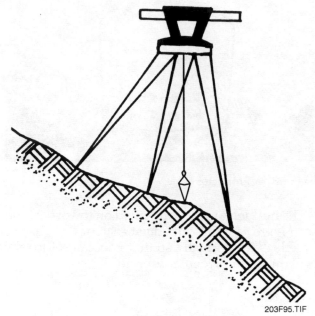

Figure 95 ◆ Leveling a tripod using a plumb bob.

203F96.EPS

Figure 96 ◆ Stadia rod.

9.5.1 Setting Up the Builder's Level

To set up the builder's level, determine the proper location and force the tripod legs into the earth approximately 3 feet apart. Check the leveling vial to determine whether the legs should be adjusted. This is a rough measurement. With the tripod legs firmly positioned, you can then accurately level the instrument.

Rotate the telescope until it lies over two opposing leveling screws. Move the two screws in opposite directions until the bubble is centered in the leveling vial. After rotating the telescope 90 degrees, adjust the two remaining screws. To ensure accuracy, repeat the leveling procedure at least one more time. With the builder's level properly set up, you're ready to measure grade.

9.5.2 Measuring Grade With the Builder's Level

For plumbing jobs, you often will measure elevations between two points that are below the line of sight of the builder's level. This is a two-person job—one at the telescope, one at the stadia rod (see *Figure 96*). Position the builder's level midway between the two points and set it up as discussed in the previous section.

A movable target on the stadia rod provides accuracy in reading the measurements. The stadia rod may be marked off in feet, feet and inches, tenths of a foot, or hundredths of a foot. You may substitute a folding rule for the stadia rod.

To convert decimals of a foot to the equivalent inches, refer to *Table 3*. Conversions are shown for measurements ranging from ⅛ inch to 3 inches.

Table 3 Inches Converted to Decimals of a Foot

Inches	Decimal of a Foot	Inches	Decimal of a Foot	Inches	Decimal of a Foot
⅛	.0104	3⅛	.2604	6¼	.5208
¼	.0208	3¼	.2708	6½	.5417
⅜	.0313	3⅜	.2813	6¾	.5625
½	.0417	3½	.2917	7	.5833
⅝	.0521	3⅝	.3025	7¼	.6042
¾	.0625	3¾	.3121	7½	.6250
⅞	.0729	3⅞	.3229	7¾	.6458
1	.0833	4	.3333	8	.6667
1⅛	.0938	4⅛	.3438	8¼	.6875
1¼	.1042	4¼	.3542	8½	.7083
1⅜	.1146	4⅜	.3646	8¾	.7292
1½	.1250	4½	.3750	9	.7500
1⅝	.1354	4⅝	.3854	9¼	.7708
1¾	.1458	4¾	.3958	9½	.7917
1⅞	.1563	4⅞	.4063	9¾	.8125
2	.1667	5	.4167	10	.8333
2⅛	.1771	5⅛	.4271	10¼	.8542
2¼	.1875	5¼	.4375	10½	.8750
2⅜	.1979	5⅜	.4479	10¾	.8958
2½	.2083	5½	.4583	11	.9167
2⅝	.2188	5⅝	.4688	11¼	.9375
2¾	.2292	5¾	.4792	11½	.9583
2⅞	.2396	5⅞	.4896	11¾	.9792
3	.2500	6	.5000	12	1.0000

Example: 4⅜ inches is 0.3646 of a foot. 90° 20¼"

Now you are ready to measure the elevations. Follow these steps, referring to *Figure 97*:

Step 1 Hold the stadia rod on top of the first point.

Step 2 Look through the builder's level and focus on the stadia rod.

Step 3 Signal the person holding the stadia rod to move the target up or down until it is centered in the telescope's cross hairs. (The vertical cross hair is used to make sure the stadia rod is held vertical.)

Step 4 Read the numbers at the horizontal cross hairs to measure the elevation (see *Figure 98*).

Step 5 Repeat this procedure to obtain the relative elevation of the second point.

Step 6 Once you have the relative elevation and the distance between both points, use the grade formula to calculate the grade.

Occasionally, you will have to determine the elevation for lines of pipe that are to be run above the height of the builder's level (see *Figure 99*). This is also a two-person job. To measure this elevation, follow these steps:

Step 1 Position the builder's level between the two unknown points and set it up.

Step 2 Have your partner hold the stadia rod *upside down* against the desired point.

Step 3 Sight through the telescope and take a reading.

Step 4 Repeat this procedure to obtain the elevation of the second point.

The cold beam laser has recently come into wide use in the plumbing industry (see *Figure 100*). It projects a fine beam of infrared light that is perfectly straight and will not sag. Also, the beam does not expand appreciably. Grade set with the cold beam laser is highly accurate. Although the laser is considered harmless, *never* look directly into the beam of light.

You may position the laser inside the pipe, outside the pipe, or even at grade level above the pipe (see *Figure 101*). Use a tilting telescope to determine the height of the beam below ground. To adjust the laser unit, shoot a reading on a stadia rod and transfer it down to the laser unit.

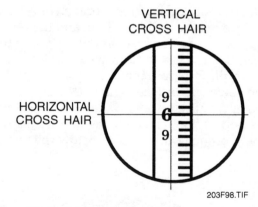

Figure 97 ◆ Measuring elevations.

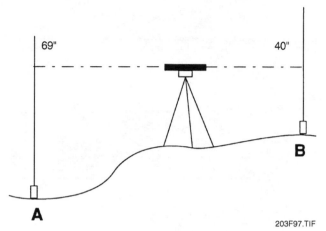

Figure 98 ◆ Reading the elevation.

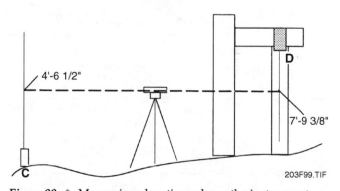

Figure 99 ◆ Measuring elevations above the instrument.

ON THE LEVEL

Using a Calculator to Convert Decimals

You can use a calculator to convert decimals of a foot to the equivalent inches. Simply multiply by 12. For example, if you want to convert 0.0833 decimals of a foot into inches, multiply by 12. The answer is 0.99 or 1. On the other hand, if you want to convert inches into decimals of a foot, divide by 12 (1 divided by 12 is 0.0833). Try this yourself. Convert 2 inches into decimals of a foot using your calculator. (The answer is 0.1667.)

Figure 100 ◆ Cold beam laser.

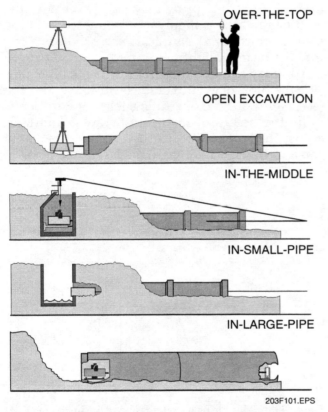

OVER-THE-TOP

OPEN EXCAVATION

IN-THE-MIDDLE

IN-SMALL-PIPE

IN-LARGE-PIPE

203F101.EPS

Figure 101 ◆ Positioning the laser.

To position a target, place it inside the pipe, or hang it from the outside (see *Figure 102*). Turn the cold beam laser on. To set the pipe at the correct grade, move the end of the pipe until the laser beam strikes the target in its center (see *Figure 103*).

Although the method of calculating grade is the same in all cases, you must be able to run pipe at a specific grade for many different situations. The next section will discuss computing grade for several common plumbing jobs.

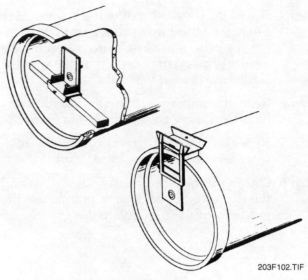

Figure 102 ◆ Positioning the target.

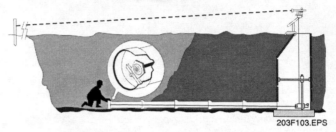

Figure 103 ◆ Positioning the pipe.

9.5.3 Computing Grade for Unobstructed Runs of Piping

The grade for roof drains, fixture branch lines, and occasionally even the building drain and the sewer main itself can be determined simply by using the grade formula. In this example, we'll compute the grade for a building drain that runs from the building to the sewer main (see *Figure 104*). Using the stadia rod and builder's level, take a reading at the first elevation of 9'-6". This means that the top of the building branch outlet is 9'-6" below the line of

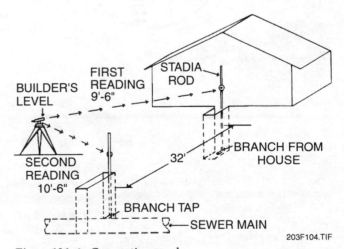

Figure 104 ◆ Computing grade.

sight established through the telescope on the builder's level. Repeating the sighting procedure, take a reading at the second elevation of 10'-6". With both points established, you can compute the grade. The difference in elevation, or fall, is 12 inches (10'-6" – 9'-6" = 12"). Remember that for the grade formula, fall is always expressed in inches. (As discussed previously, grade equals fall divided by run.) In *Figure 104*, you'll see that the run is 32 feet. Therefore, the grade is ⅜ inch per foot.

$$G = \frac{F}{R}$$

$$G = 12"/32" = 0.375 = ⅜" \text{ per foot}$$

9.5.4 Computing Grade Using Batter Boards

You can also use the builder's level to accurately position **batter boards** (see *Figure 105*). To use batter boards to measure the grade of a line of pipe, follow these steps:

Step 1 Drive a stake on either side of the excavation trench.

Step 2 Drive a second stake on the opposite side of the trench, in line with the first stake. (If a length of pipe has already been positioned in the trench, drive the stakes beside the hub end of the pipe.)

Step 3 Place the batter board in position and level it using a builder's level.

Step 4 For ease in calculating grade, locate the batter board at a convenient height above the pipe. (You can use the general purpose level for this task, although it is not as accurate as the builder's level.)

Step 5 Once the batter board is correctly positioned, use C-clamps or nails to attach it to the stakes.

Step 6 Use a plumb bob or level to transfer the centerline of the run of pipe to the batter board.

Step 7 Drive a nail into the batter board at the height established in Step 4.

Step 8 Repeat this procedure every 25 feet to 50 feet.

Step 9 Stretch a string line (which becomes the line of reference) tightly between the nails on the batter boards.

Step 10 Measure from the string down to the pipe as you lay the pipe.

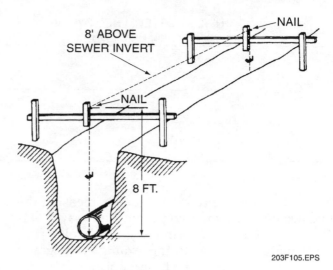

203F105.EPS

Figure 105 ◆ Positioning the batter boards.

9.5.5 Computing Grade for a Roof Drain Within the Ceiling

Roof drains, placed according to local code requirements, are located at various points on flat-roofed structures to remove rainwater. These drains are connected to a storm drainage stack.

Problems may arise when obstacles occur between the roof drain and the stack. Such obstacles may be roof trusses, heating and ventilation ducts, electrical conduit, and other plumbing lines. These obstacles require you to carefully determine the grade of the branch line so that its grade can be constant.

A typical problem is illustrated in *Figure 106*. A duct and a pipe are located between the roof drain and the drainage stack. The slope of a line that will pass both under the duct and over the pipe is the desired grade for the branch line. To measure that grade, you must measure down from the duct the radius of the pipe. You must also add clearance space.

For example, let's assume that a 4-inch pipe is to be used for the branch line. The pipe must pass at least 2 inches (the radius of the pipe) under the duct. You'll need to figure in the thickness of the pipe wall as well. Let's also assume that you want a minimum of 1-inch clearance. Therefore, the center of the branch line must pass approximately 3 inches under the duct. Using the floor of the crawl space as a reference, measure up to a point within 3 inches of the duct. Let's say this distance is 15 inches. You must determine the relative elevation of the second critical point. That is the point at which the branch line will pass above the obstacle pipe. In the same way as before, you determine that the centerline of the branch line must pass at a relative elevation of 13.5 inches from the obstacle pipe. The elevations are relative to the floor of the crawl space.

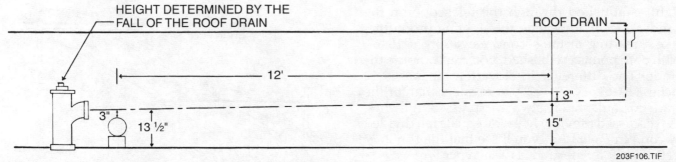

HEIGHT DETERMINED BY THE
FALL OF THE ROOF DRAIN

ROOF DRAIN

12'

13 ½"   3"

3"

15"

203F106.TIF

Figure 106 ◆ Figuring grade.

To calculate the grade for the line, measure the distance between the two points. Let's say this is 12 feet. Use the grade formula (grade = fall ÷ run) to calculate the grade. In this example, the grade for the branch line is ⅛ inch per foot.

$$G = \frac{F}{R}$$

$$G = 1.5"/12' = 0.125" = \frac{1}{8}" \text{ per foot}$$

Note that the height of the fitting used to connect the branch line to the stack and the height of the bend used to direct the branch line to the roof drain will both be determined by these measurements.

9.5.6 Computing Grade for Hanging Pipe Trapezes

Pipe run near the ceiling of a structure is generally supported with a pipe trapeze (see *Figure 107*). Local codes specify how far apart trapezes may be set. When drainage and waste pipe is involved, you must install the trapezes so that they allow the pipe to slope toward the direction of flow. Therefore, each trapeze must be installed at a lower height (farther from the ceiling) than the previous trapeze.

Install the first trapeze at an appropriate distance from the ceiling. That distance should be enough to allow room for the pipe and for you to work. After the trapeze is installed, level it using a torpedo level or other leveling device.

Figure 107 ◆ Pipe trapeze.

203F107.TIF

Locate and set up a builder's level beside the desired path of the pipe (see *Figure 108*). Place an extended folding rule upside down against the installed trapeze. Sight through the builder's level and take a reading. Let's assume the reading is 8'-2". This reading is relative and tells you nothing in itself. However, it does provide you with a starting point.

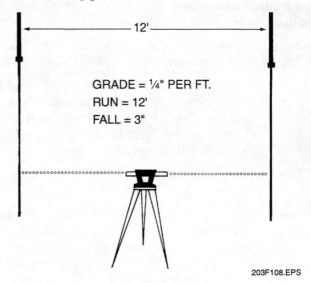

12'

GRADE = ¼" PER FT.
RUN = 12'
FALL = 3"

203F108.EPS

Figure 108 ◆ Determining grade.

For example, assume that you want a grade of ¼ inch per foot. Also assume the trapezes will be positioned 12 feet apart. Use the grade formula to determine that the line of pipe will fall 3 inches over the 12-foot distance between the two adjacent trapezes.

$$F = G \times R$$

$$F = \frac{1}{4}" \times 12' = 3"$$

By knowing this fall and the relative elevation of the first trapeze, you can determine the elevation of and adjust the second trapeze. If the first trapeze was located 8'-2" from the ceiling and the fall of the pipe to the next trapeze is 3 inches, then the second trapeze must be set at a relative elevation of 7'-11" (8'-2" − 3" = 7'-11"). You will note that as the trapezes fall further downward, the subsequent readings will be less and less.

Once you know its proper relative elevation, you can install the second trapeze. Hold the bottom of the folding rule against the trapeze. Adjust the trapeze until you get a reading of 7'-11". Then, secure the trapeze in position. Repeat this procedure to install the rest of the system.

10.0.0 ◆ MODIFYING THE STRUCTURAL MEMBERS

Occasionally, structural members lie in the path of a run of pipe. Sometimes the space available is not sufficient to allow pipe to be run. In such cases, plumbers must modify structural members. This section deals with the five most common methods of modifying structural members:

- Drilling
- Notching
- Boxing
- Furring
- Building a chase

When modifying structural members, you must ensure that the structure is not weakened. If questions concerning the safety and strength of the structure arise, you should consult the local code or the local plumbing inspector.

10.1.0 Drilling

Floor joists may be drilled to allow passage for a run of pipe. A hole no larger than one-third the width of the floor joist may be drilled in its center section (see *Figure 109*). Thus, the largest hole that can be drilled in the center of a 2 × 10 floor joist would be 3⅗₁₆ inches in diameter. Furthermore, the hole must be drilled in the middle of the joist, rather than through the upper or lower third of the joist. Only one such hole is permissible.

Drilling several adjacent floor joists significantly weakens the structure.

The maximum size hole that may be drilled in the end sections of the joists is 2 inches in diameter. The hole must be at least 2 inches from the edges of the board. No holes may be drilled 6 inches from either end of the floor joist.

Note that the holes drilled in the floor joists must be larger than the pipe that passes through them to provide clearance for the pipe. Generally, you should drill a 3-inch hole for a 2-inch pipe, and a 2-inch hole for a 1½-inch pipe.

10.2.0 Notching

You may also notch structural members to provide clearance for plumbing pipes and fittings. As with drilling, local codes determine the amount and type of notching permissible. Because notching presents a greater threat to a building's structural strength than drilling, notching should be done only when absolutely necessary.

Notching is not permitted in the center one-third portion of structural members (see *Figure 110*). Therefore, notching can safely be done only near the ends of the floor joists. Also, the depth of the notch cannot exceed one-fourth of the width of the board. In cases where you are forced to notch structural members, you should use bracing. In most cases, the braces consist of short lengths of 2 × 4s nailed to both sides of the notched member (see *Figure 111*). Bracing significantly strengthens the member.

10.3.0 Boxing Floor Joists

Where either drilling or notching will not provide adequate plumbing clearance, plumbers use the boxing method. Boxing floor joists is commonly necessary where a floor joist interferes with the

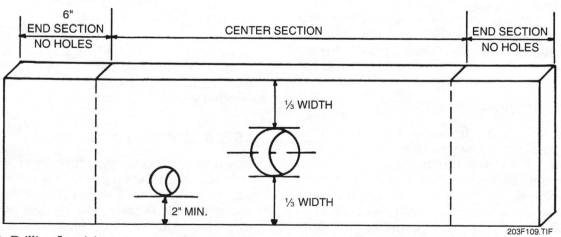

Figure 109 ◆ Drilling floor joists.

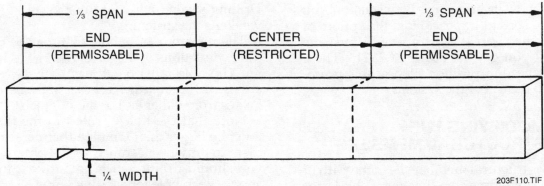

Figure 110 ◆ Notching structural members.

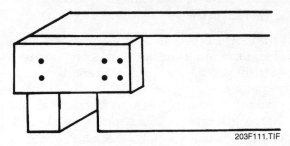

Figure 111 ◆ Bracing structural members.

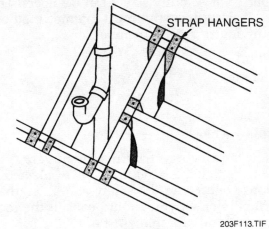

Figure 113 ◆ Using strap hangers.

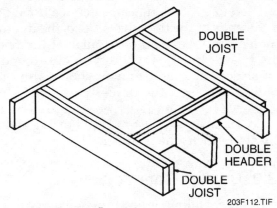

Figure 112 ◆ Boxing floor joists.

installation of the closet flange (see *Figure 112*). In this case, cut the end of the floor joist back to clear the closet bend and flange. Then, construct a double header to bridge the gap between the two adjacent floor joists and nail it into place. Using strap hangers is another method of fastening the floor joists to the header (see *Figure 113*). To further strengthen the floor, double the adjacent joists.

10.4.0 Furring Strips

The space inside the framed wall must be wide enough to allow passage of the DWV stack. In most cases, 2 × 4 stud framing does not provide sufficient space. Therefore, 2 × 6 or 2 × 8 studs are usually placed in walls where the DWV stack is located. In cases where the wall has previously been framed with 2 × 4 studs, add furring strips (see *Figure 114*). This procedure provides the

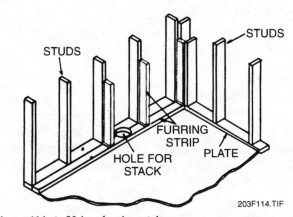

Figure 114 ◆ Using furring strips.

space required for the DWV stack. When you add furring strips, remember that you must locate the fixtures from the furred wall and not from the original wall.

10.5.0 Building a Chase

In some instances, even furring strips will not provide the necessary clearance for plumbing pipes and fixtures. This happens when a line of fixtures, such as water closets, are placed along one wall. In these situations, build a chase (see *Figure 115*). The chase is an extremely small, sealed space that provides the necessary clearance for pipes, fittings, carriers, and other pieces of plumbing. The chase may or may not extend to the ceiling and adjacent walls. Usually, it is permanently sealed, so you must properly install the plumbing before the wall finishing material is applied. Make cleanouts accessible with removable cover plates flush-mounted on the chase wall.

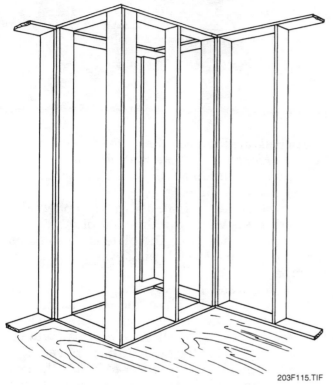

Figure 115 ◆ Building the chase.

11.0.0 ◆ TESTING AND INSPECTING DWV PIPING

In a DWV system, even a small leak can be dangerous. Incorrectly installed systems can result in explosions, toxic poisoning, and disease. These dangers become even more significant in densely populated areas. Plumbers are responsible for installing the DWV system properly and for arranging for proper tests and inspections.

Tests are part of the plumbing inspection. Inspections must be made on the following:

• All new work
• Any part of an existing system that is changed
• Any part of an existing system that is affected by new work or changes in other parts of the system

The final requirement may not apply where plumbing is placed in an outhouse, stable, or detached, uninhabited building.

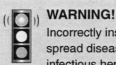

WARNING!
Incorrectly installed DWV systems can cause and spread diseases such as typhoid, scarlet fever, infectious hepatitis, poliomyelitis, and various types of dysentery.

The inspection ensures compliance with code requirements. It also ensures that the installation and construction of the finished system are the same as described in the approved plans.

11.1.0 Testing Process and the Plumbing Code

Most codes require two tests of the DWV system. The first is conducted after the plumbing system is roughed-in. Testing of the rough-in may be completed in two steps: (1) testing all parts of the DWV that will be underground and (2) testing the DWV piping that is aboveground. The second test is conducted after the fixtures are set to locate leaky traps or other unsatisfactory conditions.

Don't backfill any part of the roughed-in plumbing until it is inspected and approved. This action is prohibited and usually carries a penalty.

11.1.1 Scheduling Tests and Inspections

The person doing the work authorized by the work permit must notify the plumbing inspector that the work is ready for inspection. The inspector may require that every request for inspection be filed a certain number of working days before the inspection is wanted. These requests can be made in writing or by telephone.

11.1.2 Setting Up the Tests

If you request the tests, you must provide access to the work and the means for proper inspection. You must furnish all equipment, material, and labor necessary for inspections or tests.

11.1.3 Conducting the Tests

The required tests must be done in the presence of a plumbing inspector. You should conduct the pressure tests before the inspector arrives to ensure that the system is tight. If the plumbing is not completed or will not pass the required tests, you will have to reschedule the inspection. The best method is to complete the installation, perform a preliminary test, and then work on something else until the inspector arrives to witness the formal test. In many areas, you may be allowed to keep the pressure in the system until the inspector arrives. This procedure saves time for both the inspector and the contractor.

11.1.4 Certification and Revocation

The inspector approves the plumbing system, certifies that it complies with the local code, and posts the certificate in a conspicuous place. These certificates can be revoked for any violation of the prevailing code. This means that it is illegal for anyone to make changes in the installed and certified piping system that would violate the code. Even in areas where no plumbing code exists, an inspection may be required to satisfy the agency that finances the construction. This is certainly true for all homes financed with loans insured by the FHA (Federal Housing Administration) or the VA (Veterans Administration).

11.1.5 When Tests Show Defects

If tests show defects, the defective work or material must be replaced and the test repeated. In all cases the inspector will designate the points at which the pressure is relieved or drawn off.

11.1.6 Exceptions to Tests

Tests are *not* required after:

- Repairing or replacing old fixtures, fittings, faucets, or valves
- Forcing out stoppage
- Repairing leaks
- Relieving frozen pipes and fittings

These four cases are the most common exceptions to the requirement for testing and inspection. In addition, no tests or inspections are required where building or drainage systems are set up solely for exhibition purposes.

Repairs are limited to replacing defective components in a piping system. Installing new vertical or horizontal lines of soil, waste, vent, or interior leader pipes is considered new work. Even the change of relative location of these pipes qualifies as new work. In cases of buildings condemned because of unsanitary conditions in the plumbing, the code for new plumbing must be followed.

11.2.0 Tools and Equipment

The tools and equipment you will use to conduct the test include test plugs. Two types of test plugs are illustrated in *Figure 116*. Mechanical test plugs block fitting openings by expanding a rubber gasket inside the fitting.

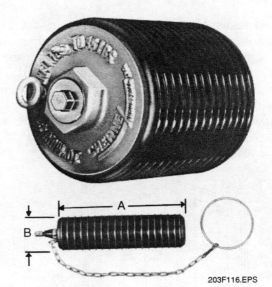

203F116.EPS

Figure 116 ◆ Air test plugs.

Test balls are inflated with air to close the opening (see *Figure 117*). Note that you can use the extension hose to inflate the test balls, which are placed inside the pipe.

The water removal plate allows a test plug to be deflated so the water can flow out of the piping system (see *Figure 118*). The plate is installed as shown to prevent the test plug from going down the pipe.

Also, the water removal plate retains most of the water within the pipe. The test manifold is connected between the pump and the piping system (see *Figure 119*). It allows the pressure to be held in the pipe and the pump to be removed. The test pump enables the plumber to inflate the test plugs and to apply the required pressure to the piping system (see *Figure 120*). The pump is equipped with a check valve and a gauge.

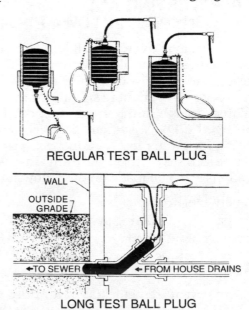

REGULAR TEST BALL PLUG

LONG TEST BALL PLUG

203F117.TIF

Figure 117 ◆ Test plug installation.

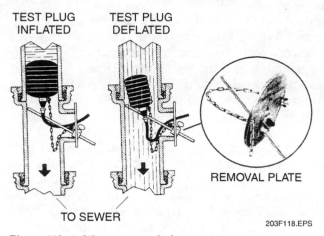

TEST PLUG INFLATED   TEST PLUG DEFLATED

REMOVAL PLATE

TO SEWER

203F118.EPS

Figure 118 ◆ Water removal plate.

203F120.EPS

Figure 120 ◆ Test pump.

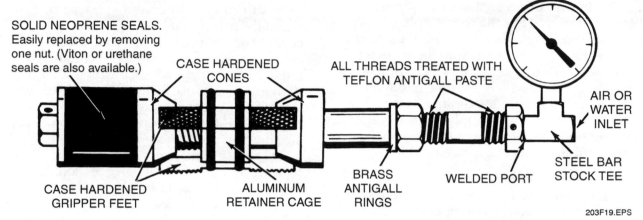

SOLID NEOPRENE SEALS. Easily replaced by removing one nut. (Viton or urethane seals are also available.)

CASE HARDENED CONES

ALL THREADS TREATED WITH TEFLON ANTIGALL PASTE

AIR OR WATER INLET

CASE HARDENED GRIPPER FEET

ALUMINUM RETAINER CAGE

BRASS ANTIGALL RINGS

WELDED PORT

STEEL BAR STOCK TEE

203F19.EPS

Figure 119 ◆ Test manifold.

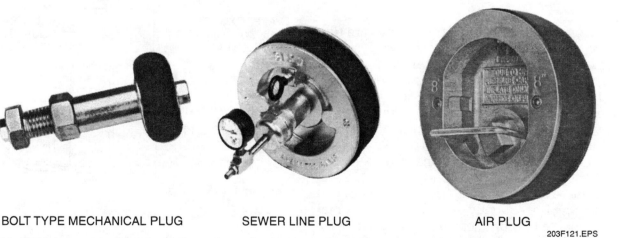

BOLT TYPE MECHANICAL PLUG

SEWER LINE PLUG

AIR PLUG

203F121.EPS

Figure 121 ◆ Common air and mechanical test plugs.

11.3.0 Testing Procedures

Four different kinds of tests are used with DWV piping:

- Water test
- Air test
- Smoke test
- Odor test

The water and air tests are the most frequently used. The other tests are used mainly to locate leaks that are difficult to find.

11.3.1 The Water Test

In the water test, all openings in the drainage pipes are closed except those at the tops of the stacks. The openings can be closed by test plugs (see *Figure 121*) or by pipe fittings such as caps or plugs.

Where test pressures are less than about 5 pounds per square inch (psi) and the pipes are large (for example, vitrified-clay or concrete pipes), other types of plugs are used. A bag of sand or clay around which **oakum** or similar material is lightly caulked is one alternative. A wooden diaphragm surrounded by an inflated inner tube and supported by a sandbag is a second alternative. Some companies produce a special rubber diaphragm with flaring edges to seal large pipes. Quick-setting plaster of paris and Portland cement formed in the pipe can serve as a plug. When a test is to be made on a waste pipe that is already connected to the sewer main, the house drain can be plugged by inserting a wye-test plug through a cleanout opening. If no other alternative is available, a section of the pipe can be broken out to permit installation of a plug.

When all openings (except at the tops of the stacks) are closed, water is run into the pipes until it overflows from the top of a stack. The water level may drop immediately after the initial filling because of air pockets in the piping system. This does not indicate a leak. Simply refill the stack until the air pockets are eliminated. The water can be introduced in one of three ways:

- Into the top of a stack with a hose
- Through a connection temporarily installed in a cleanout or other convenient opening
- Through the hollow handle of a mechanical test plug

If the water level drops after the pipes are full, there is a leak. During a satisfactory test, the water level in the pipe should remain stationary for not less than 15 minutes. All parts of the system should be subjected to a pressure of at least 10 feet of water (4.34 pounds per square inch gauge pressure [psig]). It is not desirable to use pressures over 30 to 40 feet of water. Higher stacks should be tested in sections of 10 to 40 feet. In high-rise buildings this piping is tested in sections, one story at a time. Sectional testing of DWV piping is frequently required for horizontal piping installed in commercial buildings. In these cases, the horizontal piping is sectioned off by adding test tees. The inspector will be required to witness the test of each individual section of the piping system.

11.3.2 The Air Test

In the air test, all openings are closed and an air pressure of at least 5 psi is exerted in the pipes for at least 15 minutes. Falling pressure, as shown by a sensitive gauge attached to the pipes, indicates a

leak. A large air leak can usually be detected by sound. Where little or no sound is made and you suspect a leak, you can find its location by applying a smoke or odor test to the piping system. Air tests are useful in cold weather when a water test could freeze and damage the pipes. The air test also allows you to maintain uniform pressure throughout the section being tested. However, the air test is not as simple to apply as the water test, and discovering leaks is sometimes more difficult.

The test pressure for an air test plug using a manometer or test can is about 1 inch of water column (see *Figure 122*). If, after all openings have been closed, this pressure can be maintained for 15 minutes without additional pumping, the system can be considered airtight. You may have some difficulty in interpreting the pressure readings if the pressure falls slowly because the air has cooled and not because of a leak. When in doubt, restore the air pressure to the original level without releasing air already in the pipes. If the rate of decrease of pressure is slower, the loss of pressure is probably caused by temperature change and not a leak. If the test shows that the pipes are not airtight, you must locate the leak.

To locate a leak, spray or squeeze a solution of liquid soap onto the joints in question. If a steady flow of air bubbles up through the soap solution, you have found the leak.

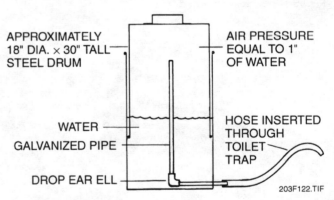

Figure 122 ◆ Test can.

11.3.3 The Smoke Test

In the smoke test, oily waste, tar paper, or similar materials are burned to produce a thick smoke that is blown into the DWV piping (see *Figure 123*). Smoke bombs, which look like large firecrackers, can be dropped inside the stack to produce a large volume of brightly colored smoke. This procedure is much simpler than using a smoke generator. Do not create smoke by mixing chemicals such as ammonia and muriatic acid. This method produces toxic smoke and dangerous particles that are hard to see.

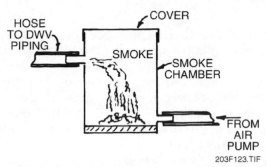

Figure 123 ◆ Smoke chamber.

When the smoke begins to escape from the top of the stack, close the stack with a test plug and increase the pressure to 1 inch of water. This pressure should be maintained until the end of the test. If the pressure cannot be maintained, a leak is indicated. To find it, look for the escaping smoke. Since the leaks are sometimes too small for the escaping smoke to be seen, close windows and doors to retain the odor of the escaping smoke and locate the leak by scent.

 DID YOU KNOW?

A World's Fair and an Outbreak of Dysentery

In 1933 during the World's Fair in Chicago, Illinois, faulty plumbing in two hotels resulted in an outbreak of dysentery. Of 1,409 reported cases, 98 people died. When speaking about the tragedy in 1934, Major Joel Connolly of the Chicago Bureau of Sanitary Engineering said, "... plumbing demands the very best painstaking effort that thoroughly qualified, certified plumbers can give ... especially where large numbers of people may be affected by the contamination of water."

11.3.4 The Odor Test

You can use the odor test if other tests fail to locate the leak. Use oil of peppermint or ether to create the odor. Close the outlet end of the drainage system and all vent openings except the top of one stack. Empty about one ounce of oil of peppermint for every 25 feet of stack (but not less than two ounces) down the stack. Pour a gallon or more of warm water into the stack and immediately close the top of the stack.

Add air pressure to force the odor through the leak. Search for the leak using your sense of smell. Anyone who has recently handled peppermint should not enter the building until the search is complete. The odor test is simpler than the smoke test, but results are not always satisfactory because there may have been insufficient pressure in the pipes to force the odor through a leak. Also, it may be difficult to locate the leak after you've detected the odor.

Review Questions

Sections 7.0.0–11.3.4

1. The best way to secure the position of DWV piping in slab-on-grade construction is to _____.
   a. use tie wire to secure pipe to two perpendicular pieces of rebar
   b. place sleeves around pipes that pass through footings and walls
   c. bore holes in the floor for DWV pipes after the concrete cures
   d. mound excavated dirt around the piping to keep it in place

2. Metal straps on plastic DWV pipe _____.
   a. are recommended for all residential DWV systems
   b. cannot be attached with powder actuated fastening systems
   c. are restricted to single-family homes by most building codes
   d. do not allow for thermal expansion and contraction of the pipe

3. The formula for calculating the grade in a DWV piping system is _____.
   a. Grade = Fall ÷ Run
   b. Grade = Run ÷ Fall
   c. Grade = Fall × Run
   d. Grade = Elevation × Fall ÷ Run

4. Notching structural members to provide clearance for plumbing pipes is _____.
   a. permitted anywhere along a supporting member
   b. restricted to the center one-third portion of supporting members with bracing
   c. limited to no more than one-half the width of the supporting member
   d. allowed only near the ends of supporting members with bracing

5. The _____ is most commonly used to test DWV piping for leaks.
   a. air test
   b. odor test
   c. water test
   d. smoke test

Summary

Experienced plumbers develop an action plan for a DWV installation that includes ordering, delivery, and storage of fixtures and materials; plans and specifications; schedules; slope and grade calculations; testing and inspection; safety; and accessibility requirements.

When installing DWV systems, plumbers pay attention to both on-the-job safety and the health and safety of the people who will eventually use these systems. While plumbers can find installation requirements in plans, prints, or local codes, they also rely on their experience and skill to complete the installation and solve problems on site.

Included in the plumber's responsibilities are locating fixtures, hanging and supporting pipes, installing the stack, calculating grade, modifying structural members, testing the installation, and arranging for the inspection. Plumbers do all of these jobs safely and accurately to ensure that the piping system does its job of delivering clean water to the structure, moving wastes away from the structure, and keeping harmful gases and disease out.

Trade Terms Introduced in This Module

Accessibility requirements: The requirements outlined in some building codes and by ANSI (the American National Standards Institute) that affect physically challenged people and their access to public buildings and facilities.

Batter board: One of a pair of horizontal boards that are nailed at right angles to each other to three posts set beyond the corners of a building elevation. Strings fastened to these boards indicate the exact corner of a building.

Bench mark: In surveying, a marked reference point on a permanent, fixed object such as a metal disk set in concrete, the elevation of which is known, and from which the elevation of other points may be obtained.

Blocking: Pieces of wood used to secure, join, or reinforce members, or to fill spaces between them.

Carrier fittings: The support apparatus for wall-hung bathroom fixtures.

Chase: The plumbing access area between two restrooms.

Cure: A natural process in which excess moisture in concrete evaporates and the concrete hardens.

I-beam: A rolled or extruded structural metal beam having a cross-section that looks like the letter I.

Invert elevation: In plumbing, the lowest point or the lowest inside surface of a channel, conduit, drain, pipe, or sewer pipe. It is the level at which fluid flows.

Manufacturers' specification sheets: Information from fixture manufacturers that shows the rough-in dimensions for fixture models that they sell.

Material takeoff: A list of the type and quantity of material required for a given job.

Oakum: A loose, stringy fiber (created from taking apart old ropes) that is treated with tar and used as a caulking material.

Rebar: A ribbed steel bar that provides a good bond when used as a reinforcing bar in concrete.

Sewer tap: The inlet of a private sewage disposal system.

Sheathing: The covering (usually over wood boards, plywood, or wallboard) that is placed over the exterior framing or rafters of a building. It provides a base for the application of exterior cladding.

Sheetrock: A proprietary name for gypsum board. Also called *wallboard*.

Shoring: The act of using timbers set diagonally to temporarily hold up a wall. In excavations, the use of such timbers to stabilize the ditch sides is employed to prevent cave-ins.

Slab-on-grade: A term describing a building in which the base slab is placed directly on grade without a basement.

Sleeve: A piece of plastic piping through which water supply or drain, waste, and vent piping is inserted when that piping penetrates a building's structural elements such as concrete footings or floors. The sleeve protects the piping from being damaged by concrete.

Slope: The ground elevation or level planned for or existing at the outside walls of a building or elsewhere on the building site.

Soleplate: A horizontal timber that serves as a base for the studs in a stud partition.

Spirit level: A level that contains a small, transparent container that is partially filled with a fluid, or spirit, and has a small air space. When the level is horizontal, the air space appears as a bubble. If the vial is placed on a perfectly level surface, the bubble is centered.

Spirit vial: A small, transparent container that is partially filled with a fluid, or spirit. It is the basic component of leveling tools.

Time-critical material: Material that must arrive at a job site at a specified time and that must be installed before any other work can proceed.

Top plate: The top horizontal member of a frame building to which the rafters are fastened.

Vernier: An auxiliary scale that slides against, and is used in, reading a primary scale. The scale makes it possible to read a primary scale much closer than one division of that scale.

Answers to Review Questions

Sections 2.0.0–3.3.0
1. b
2. d
3. c
4. c
5. a

Sections 4.0.0–4.8.0
1. b
2. a
3. c
4. a
5. d

Sections 5.0.0–6.8.0
1. a
2. b
3. c
4. a
5. c

Sections 7.0.0–11.3.4
1. a
2. d
3. b
4. d
5. c

Figure Credits

Big Boy Products, Inc. 203F120

Caddy Fasteners, 203F40
Division of Van Huffel
Tube Company

Cherne Industrial Inc. 203F117, 203F118

David White 203F89, 203F92, 203F93, 203F94
Instruments

Eljer Plumbingware 203F03
Division,
Wallace-Murphy
Corporation

Expando Tools Inc. 203F119

Expando Tools, Inc. 203F121
and Sidu
Manufacturing
Company, Inc.

Fee and Mason 203F39, 203F41, 203F44, 203F45, 203F46,
Company, Division 203F47, 203F48, 203F55, 203F56, 203F57,
of ATO 203F58, 203F59, 203F62, 203F63, 203F64,
203F65

Halsey Taylor 203F17

Hilti Fastening 203F49, 203F50, 203F51, 203F52
Systems, Inc.

ITT Grinnell 203F36, 203F37, 203F60, 203F61
Corporation

Josam Manufacturing 203F09, 203F10, 203F16, 203F20, 203F21,
Company 203F23

L. S. Starrett Company 203F88

Power-Strut, Division of Van Huffel Tube Company	203F42, 203F43, 203F74, 203F75
Red Head, ITT Phillips Drill Division	203F53, 203F54
Ridgid Tool Company	203F82, 203F85, 203F86, 203F87
Sidu Manufacturing Company, Inc. and Cherne Industrial	203F116
Specialty Products Company	203F35, 203F38
Spectra-Physics	203F100, 203F101, 203F102, 203F103
Tyler Pipe	203F14, 203F15
Zurn Industries	203F05, 203F11, 203F12, 203F13, 203F19, 203F22, 203F25

NCCER CRAFT TRAINING USER UPDATES

The NCCER makes every effort to keep these textbooks up-to-date and free of technical errors. We appreciate your help in this process. If you have an idea for improving this textbook, or if you find an error, a typographical mistake, or an inaccuracy in the NCCER's Craft Training textbooks, please write us, using this form or a photocopy. Be sure to include the exact module number, page number, a detailed description, and the correction, if applicable. Your input will be brought to the attention of the Technical Review Committee. Thank you for your assistance.

Instructors – If you found that additional materials were necessary in order to teach this module effectively, please let us know so that we may include them in the Equipment and Materials list in the Instructor's Guide.

Write: Curriculum Revision and Development Department
National Center for Construction Education and Research
P.O. Box 141104, Gainesville, FL 32614-1104

Fax: 352-334-0932

E-mail: curriculum@nccer.org

Craft _____ Module Name _____

Copyright Date _____ Module Number _____ Page Number(s) _____

Description _____

_____

_____

_____

(Optional) Correction _____

_____

_____

(Optional) Your Name and Address _____

_____

_____

Installing Roof, Floor, and Area Drains

COURSE MAP

This course map shows all of the modules in the second level of the Plumbing curriculum. The suggested training order begins at the bottom and proceeds up. Skill levels increase as you advance on the course map. The local Training Program Sponsor may adjust the training order.

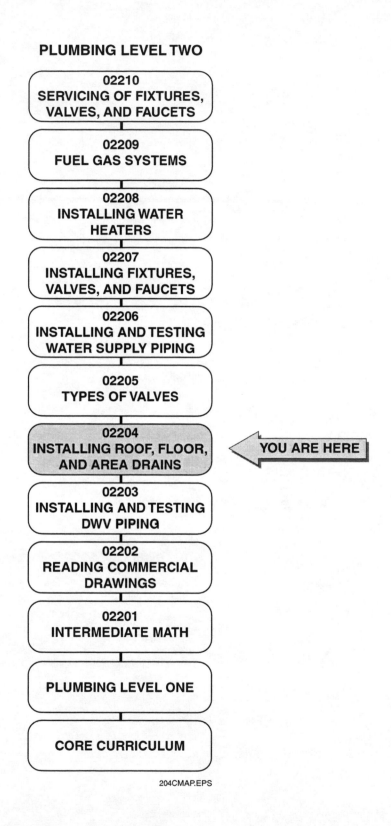

PLUMBING LEVEL TWO

02210
SERVICING OF FIXTURES, VALVES, AND FAUCETS

02209
FUEL GAS SYSTEMS

02208
INSTALLING WATER HEATERS

02207
INSTALLING FIXTURES, VALVES, AND FAUCETS

02206
INSTALLING AND TESTING WATER SUPPLY PIPING

02205
TYPES OF VALVES

02204
INSTALLING ROOF, FLOOR, AND AREA DRAINS ◁ YOU ARE HERE

02203
INSTALLING AND TESTING DWV PIPING

02202
READING COMMERCIAL DRAWINGS

02201
INTERMEDIATE MATH

PLUMBING LEVEL ONE

CORE CURRICULUM

204CMAP.EPS

MODULE 02204 CONTENTS

Figures

Tables

Installing Roof, Floor, and Area Drains

Objectives

When you have completed this module, you will be able to do the following:

1. Use a surveyor's level or transit level to set the elevation of a floor or area drain.
2. Install a roof, floor, and area drain.
3. Install waterproof membranes and flashing.

Prerequisites

Before you begin this module, it is recommended that you successfully complete the following modules: Core Curriculum; Plumbing Level One; Plumbing Level Two, Modules 02201 through 02203.

Required Trainee Materials

1. Paper and sharpened pencil
2. Appropriate Personal Protective Equipment

1.0.0 ◆ INTRODUCTION

Roof, floor, and area drains collect storm water and direct it to the storm drainage system. The piping that connects the roof and area drains to the storm sewer is separate from the drain, waste, and vent (DWV) system of the sanitary sewer. However, plumbers use the same materials to install both piping systems. The major difference is that traps generally are not required in storm drainage piping. Floor drains are connected to the sanitary sewer system because the wastes entering them usually are contaminated. These wastes must be treated before the water carrying them is returned to the natural water supply. Floor drains are included in this module because their design

and installation requirements are similar to those for roof and area drains.

In this module, you'll learn about the different types of roof, floor, and area drains and the specifics of installing each.

2.0.0 ◆ BASIC PARTS OF DRAINS

Manufacturers offer roof, floor, and area drains in several basic parts and in a wide variety of styles. You can interchange these basic parts and produce an almost endless variety of drains. The basic parts are the **drain body, grate (dome), deck clamp, clamping ring (flashing clamp), drain receiver,** and **extensions** (see *Figure 1*).

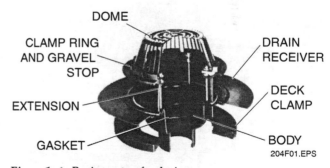

Figure 1 ◆ Basic parts of a drain.

2.1.0 Drain Body

The basic component of all drains, the drain body, funnels the water into the piping system. Usually made of cast iron or plastic, it is designed to connect to the roof, floor, or foundation. Special installations may require stainless steel or porcelain-enameled cast iron. You can see different types of standard roof drains in *Figure 2* and typical floor drains in *Figure 3*.

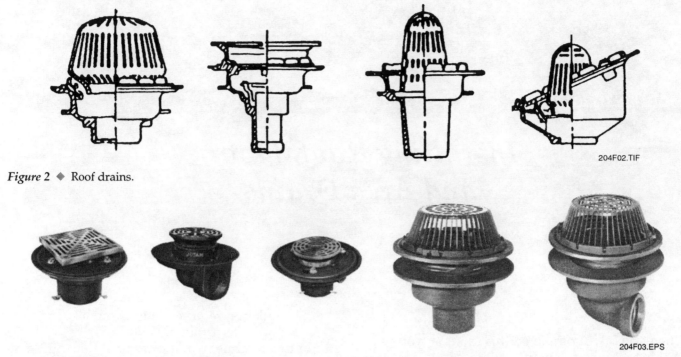

Figure 2 ◆ Roof drains.

204F02.TIF

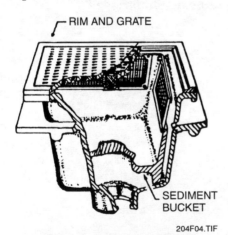

Figure 3 ◆ Floor drains.

204F03.EPS

Floor and area drains often contain **sediment buckets** to prevent solids from entering the piping (see *Figure 4*). The drain body connects to the piping system in one of several ways. Inside-caulk drain bodies are compatible with hub-and-spigot cast-iron pipe. A threaded drain body can receive a standard male pipe thread. Drain bodies can be joined with no-hub gaskets and clamps.

RIM AND GRATE

SEDIMENT BUCKET

204F04.TIF

Figure 4 ◆ Floor drain with sediment bucket.

2.2.0 Grate (Dome)

The grate, or dome, prevents debris from entering the piping system and helps reduce the chance of pipe clogs (refer to *Figures 2* and *3*). It also protects the drain by allowing people and vehicles to pass over it without damage. A roof drain with an adjustable dome controls the rate at which water

enters the piping system (see *Figure 5*). You can adjust this type of drain on site.

2.3.0 Deck Clamp

The deck clamp is a ring that fastens to the drain body with several bolts. Tightening the bolts clamps the drain securely to the roof deck (see *Figure 6*).

DOME
CLAMPING RING
MULTI-WEIR BARRIER
BAYONET-TYPE LOCKING DEVICE
GRAVEL
ROOF SUMP RECEIVER
INSULATION
WATER-PROOF MEMBRANE
METAL ROOF DECK
DECK CLAMP
EXTENSION SLEEVE
OUTLET CONNECTION

204F05.EPS

Figure 5 ◆ Adjustable roof drain dome.

ROOF DECK
DECK CLAMP

204F06.TIF

Figure 6 ◆ Deck clamp.

2.4.0 Clamping Ring (Flashing Clamp)

The clamping ring, or flashing clamp, secures the subsurface flashing or waterproof membrane to the drain body (see *Figure 7*). A roofer or plumber installs this ring when installing the flashing or waterproof membrane.

2.5.0 Drain Receiver

The drain receiver (*Figure 8*) helps support a roof drain body. It serves two important functions. First, it distributes the roof drain's weight over a large area, which is particularly important when the drain is installed over insulation. Second, it supports a drain in an oversized or off-center roof opening.

2.6.0 Extension

A drain should be flush with the roof or floor. If either the roof deck or concrete floor is too thick, use an extension to add the required height as shown in *Figure 9*. Floor drains allow you to make some adjustments without extensions because the strainers are threaded (see *Figure 10*).

3.0.0 ◆ SPECIAL DRAINS

Some structures, such as hospitals, laboratories, cafeterias, and restaurants, require a high degree of sanitation in their drains. Some installations require drains that can be washed, flushed, or primed. Special drains meet these needs.

3.1.0 Can Wash Drain

A **can wash drain** permits the user to invert a **can** over the top of the drain and wash the inside of the can (see *Figure 11*). A sediment bucket is a part of this type of drain. Can wash drains come in several sizes with either side or bottom outlets. You must connect a water supply pipe and remote valve to the can wash drain.

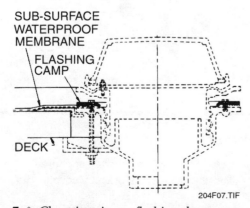

Figure 7 ◆ Clamping ring or flashing clamp.

Examples of areas where can wash drains are used include packing plants, areas where food is processed, hospitals, and any area where solids get washed into floor drains but remain in the trap. The flushing action inside the can allows the solids to pass through the trap. In nonflushing drains, solids sit, creating foul odors.

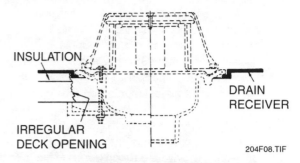

Figure 8 ◆ Drain receiver.

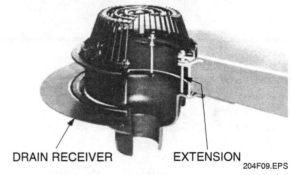

Figure 9 ◆ Extension.

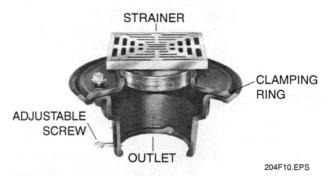

Figure 10 ◆ Threaded strainers make floor drains adjustable.

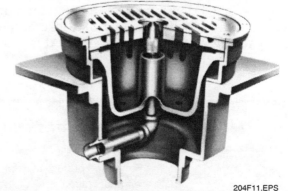

Figure 11 ◆ Can wash drain.

3.2.0 Floor Sinks

Floor sinks are used in installations that require a high degree of sanitation, such as hospitals, laboratories, cafeterias, and restaurants. The basic difference between a floor drain and a floor sink is that the floor sink is coated with a porcelain or other easily cleaned finish. Floor sinks reduce the amount of contaminant that can collect in the grate and drain body (see *Figure 12*). They come in a variety of sizes and in round, square, or rectangular shapes.

3.3.0 Flushing Floor Drain

In a **flushing floor drain** installation, an inlet is provided in the trap. Connect a water supply pipe to this inlet to flush the trap (see *Figure 13*). Be sure to fit the water supply with a valve and vacuum breaker to prevent the possibility of backflow.

Flushing floor drains are used in areas where solids get washed down into the floor drain, but remain in the trap. Examples of these areas include packing plants, areas where food is pro-

cessed, and hospitals. The flushing allows solids to pass through the trap. In nonflushing drains, solids sit, creating foul odors.

3.4.0 Trap Primer

When you expect a trap to receive limited use, install a **trap primer**, which periodically adds a small amount of water to the trap. This keeps the trap seal from failing as a result of evaporation. A deep seal trap with a trap primer connection is shown in *Figure 13*. The water supply pipe connects at the same point as the flushing connection.

Manufacturers offer a variety of **trap primer valves**. You must install some in conjunction with the water supply valve to a fixture. Each time the valve operates, a small amount of water enters the trap. Other trap primer valves meter water into the trap at regular intervals (see *Figure 14*).

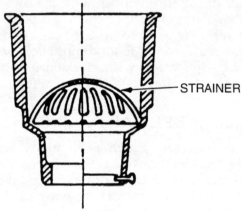

STRAINER

204F12.TIF

Figure 12 ◆ Floor sink.

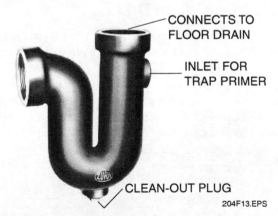

CONNECTS TO FLOOR DRAIN

INLET FOR TRAP PRIMER

CLEAN-OUT PLUG

204F13.EPS

Figure 13 ◆ Deep seal trap with primer connection.

ON THE

LEVEL

Floor Drains and Poured Concrete

Many floor drains come with a plastic cover to protect them when concrete is poured. For added protection, experienced plumbers will further protect the drain by covering it with duct tape.

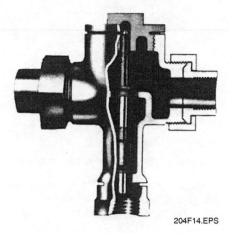

Figure 14 ◆ Trap primer valve.

204F14.EPS

4.0.0 ◆ DETERMINING REQUIREMENTS FOR FLOOR DRAINS

To determine what types of roof, floor, and area drains to install, refer to the local building code or plans and specifications. For installations that require a high degree of sanitation, contact the local health department.

4.1.0 Codes

The local code regulates small buildings and structures not designed by engineers. *Table 1* includes examples of typical code regulations. Note that the code does not state drain location and sizing. You can get this information from the drawings and specifications. If necessary, refer to the manufacturer's literature to determine the capacity of a particular drain.

4.2.0 Plans and Specifications

The plans and specifications state what types of drains are used in large construction projects such as commercial buildings, hospitals, and manufacturing facilities. These buildings have so many unique features, it is impossible to write a code that covers all the possible alternatives. So engineers custom-design the piping systems to meet specific needs and then write the plans and specifications. A partial set of specifications covering drains is shown in *Table 2*. Generally, you'll find this information in Division 15 of the specifications.

ON THE LEVEL

Floor Drain With a Funnel

This type of floor drain is used around air-conditioning units that have condensate lines. The condensate line is dropped into the funnel. For example, if a small air handler is on the second floor, the condensate line would be piped to a funnel drain on the main floor.

204SA01.EPS

Table 1 Typical Plumbing Code Regulations for Floor and Roof Drains

7.15 FLOOR DRAINS

7.15.1 Trap and Strainers. Floor drains shall have metal traps and a minimum water seal of 3 inches and shall be provided with removable strainers. The open area of a strainer shall be at least ⅔ of the cross-sectional area of the drain line to which it connects.

7.15.2 Size. Floor drains shall be of a size to efficiently serve the purpose for which they are intended.

13.5 ROOF DRAINS

13.5.1 Material. Roof drains shall be of cast iron, copper, lead, or other acceptable corrosion-resistant material.

13.5.2 Strainers. All roof areas, except those draining to hanging gutters, shall be equipped with roof drains having strainers extending not less than 4 inches above the surface of the roof immediately adjacent to the roof drains. Strainers shall have an available inlet area, above roof level, of not less than 1½ times the area of the conductor or leader to which the drain is connected.

13.5.3 Flat Decks. Roof drain strainers for use on sun decks, parking decks, and similar areas, normally serviced and maintained, may be of the flat-surface type, level with the deck and shall have an available inlet area not less than two times the area of the conductor or leader to which the drain is connected.

13.5.4 Roof Drain Flashings. The connection between roofs and roof drains that pass through the roof and into the interior of the building shall be made watertight by the use of proper flashing material.

Table 2 Specifications for Drains

P-5 Roof Drain
Shall be J.R. Smith Figure No. 1010 if no insulation above roof slab and Figure No. 1015 if roof insulation is above roof slab. Provide Figure No. 1710 expansion joint if no offset in rainwater leader line is required. Size as shown on drawings.

P-6 Floor Drain (Regular)
Shall be J.R. Smith Figure No. 2010-T, cast-iron floor drain with sediment bucket and polished nickel bronze and adjustable strainer with flashing clamp device. Size as shown on drawings. (Flashing clamp required on drains installed above first floor, slab on grade.)

P-11 Floor Drain (Equipment Rooms)
Floor drain shall be J.R. Smith Figure No. 2233 round cast-iron 12-inch diameter area drain, with sediment bucket, 4-inch pipe connections, and polished bronze top. Furnish complete with a flange. A clamp device is required if the drain is installed above first floor, slab on grade.

4.3.0 Health Department

Consult with the local health department before you install drains in places that require high levels of sanitation such as hospitals or food preparation areas. Health department inspectors must approve these installations.

> **WARNING!**
> It is up to you as plumbers and apprentices to see that all floor drains go through to the septic or sewer system by means of either a pump or gravity. Everything possible can and will go down those drains. An estimated 60 percent of farmhouses built in the 1950s and 1960s in rural America have floor drains that flow into open ditches.

Review Questions
Sections 2.0.0–4.3.0

1. The part of a floor or area drain that is designed to prevent solids from entering the piping is called the _____.
   a. extension
   b. deck clamp
   c. drain receiver
   d. sediment bucket

2. The _____ prevents debris from entering the piping system and helps reduce the chance of pipe clogs.
   a. flashing
   b. drain receiver
   c. grate or dome
   d. trap or trap inlet

3. The _____ secures the waterproof membrane to the drain body in a roof drain.
   a. clamping ring
   b. deck clamp
   c. extension
   d. grate

4. A drain that receives limited use should be installed _____ to protect against loss of the water seal.
   a. with a floor sink
   b. with a trap primer
   c. with a funnel
   d. without a trap

5. The basic component of all drains, the _____ funnels water into the piping system.
   a. drain body
   b. deck clamp
   c. extension
   d. drain receiver

5.0.0 ◆ INSTALLING FLOOR AND AREA DRAINS

Installing floor drains is similar in many ways to installing roof drains. The pipe-joining techniques are identical, correct positioning of the floor drains is critical, and you may have to attach a waterproof membrane. The steps for installing floor and area drains include locating the drains, supporting the drains, preventing damage to the drains, and installing waterproof membranes.

5.1.0 Locating Floor and Area Drains

Because the floor drain is set before the concrete floor is poured, make sure that the elevation is correct. Once the concrete has set, correcting a poor elevation is time-consuming and costly. Locate the drain at the lowest point in the floor or area. Refer to the floor plan, plumbing drawings, or specifications for the drain's position, the

> **DID YOU KNOW?**
> High humidity and moisture are associated with some biological contaminants. Places that may have high humidity levels include the following:
> - Bathrooms where humidity levels are high from bathing and showering
> - Kitchens with high humidity levels caused by cooking and dish washing
> - Laundry rooms, if the dryer is not properly vented to the outside
> - Basements with water seepage because of inadequate drainage outside the house
> - Crawl spaces with improper drainage and high levels of moisture in the soil
> - Any place with water leaks or water damage

ON THE
· LEVEL ·

Floor Drain with a Square Strainer

Floor drains with a square strainer often are chosen when the owner or engineer is working toward an overall cosmetic effect. The drains are placed in squared quarry tile to produce a more uniform look.

The general contractor usually blocks out these drains when the initial concrete is poured. The drains are later raised ¼ to ⅜ of an inch to accommodate the thicker tile.

204SA02.EPS

horizontal distances from walls, and the elevation of the finished floor drain. Coordinate your work with the concrete contractor to ensure accuracy.

It's best to make the elevation measurements with a surveyor's level or transit (see *Figure 15*). In small areas you can use a level and a straight board to locate the top of the drain with respect to a known elevation (see *Figure 16*). You can also stretch a string line across the area that indicates the highest level of the floor. Measure down from the line to determine the floor drain elevation (see *Figure 17*).

5.2.0 Supporting the Floor Drain

Supporting the drain at the proper elevation is often difficult. The area below the drain is excavated to install the pipe, so the fill is probably not stable. Use concrete blocks, bricks, or tamped gravel to support the drain body.

Sometimes you'll have to drive three or four lengths of pipe or reinforcing rod into the ground and place the drain on top of these supports. *Never* use wood blocks. They expand when concrete is poured around them and can cause the concrete to fail.

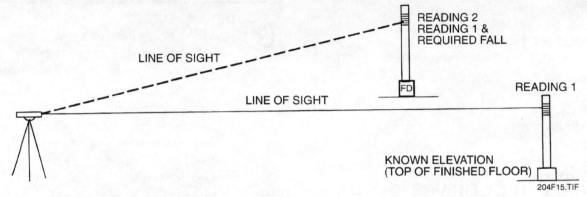

Figure 15 ◆ Measuring the elevation with a surveyor's level.

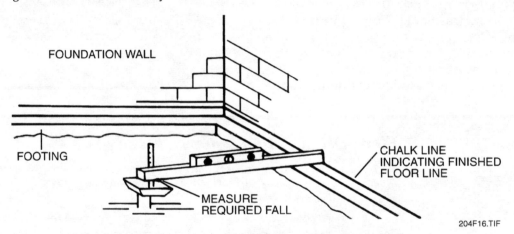

Figure 16 ◆ Using a level and straight board to locate drain elevation.

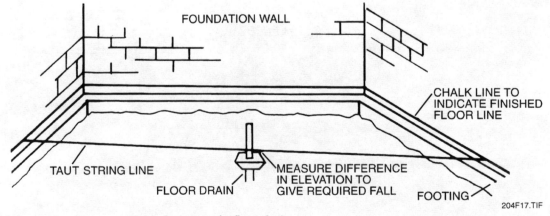

Figure 17 ◆ Stretching a string line to measure elevation of a floor drain.

Area Drains

Area drains are usually found in large exterior spaces where storm water management is a concern. These drains typically are found in small parking lots. An area drain may also be used in landscaped areas where the grade may be lower and the architect wants to avoid any water pooling. An area drain in this application will catch excess water so that during the winter months, if there is freezing and thawing, walkways will stay dry.

204SA03.EPS

5.3.0 Preventing Damage to the Floor Drain

When concrete is poured, some of it may enter the drain. To prevent this, use test plugs to temporarily block the drain. Temporarily block small drains with duct tape.

Drain grates with a special finish or polish can be damaged when cement is poured. You can remove the grate until the concrete work is completed, but be sure to protect the inside of the drain body. To do this, tape a piece of cardboard in place of the grate.

5.4.0 Installing Floor Drains With Waterproof Membranes

Some installations require a waterproof membrane between the structural concrete and the topping slab (see *Figure 18*). This type of installation is necessary for floor drains above ground level, such as showers. The membrane prevents water that may come through the topping slab from penetrating the structural concrete. The drain body is designed with weep holes so that water from the top of the membrane can enter the drain near the clamp. You must initially install the drain body without the strainer in place. After the structural concrete cures, put the waterproof membrane in place and secure it with the membrane clamp. Finally, set the strainer at the correct elevation.

If the drain is not located in the basement, you must add a vinyl or sheet lead material between the drain and the floor (see *Figure 19*). This material protects against leaks if the floor cracks.

CAUTION

Take test plugs out before installing the drain grate but after concrete is poured. Most finished drain grates are too small to allow access to retrieve the plugs.

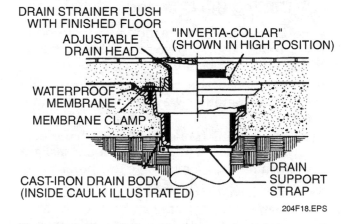

DRAIN STRAINER FLUSH
WITH FINISHED FLOOR
ADJUSTABLE
DRAIN HEAD
"INVERTA-COLLAR"
(SHOWN IN HIGH POSITION)
WATERPROOF
MEMBRANE
MEMBRANE CLAMP
CAST-IRON DRAIN BODY
(INSIDE CAULK ILLUSTRATED)
DRAIN
SUPPORT
STRAP

204F18.EPS

Figure 18 ◆ Floor drain with waterproof membrane clamping device.

204F19.EPS

Figure 19 ◆ Drain with vinyl or sheet lead material in place.

6.0.0 ◆ INSTALLING ROOF DRAINS

Installing roof drains is fairly easy, but, depending on the application and the type of drain, you will have several options to consider. The steps for installing roof drains include laying out and cutting the opening, attaching the drain body, connecting to the piping system, installing **expansion joints**, installing the waterproof membrane or flashing, and checking the drains.

6.1.0 Laying Out and Cutting the Opening

Use the dimensions given on the drawings to locate the roof drain position. Two important questions must be considered before cutting the openings:

• Is the position reasonable given the slope of the roof? Stated another way, is the location in a low point of the roof?

• Does the building frame interfere with the location you have chosen? If it does, make a slight change in the location to solve this problem.

Once you have checked the drain location, lay out the shape of the opening and cut it. The tools you use to make the cut depend upon the type of roof deck. For example, on a metal roof or decking, you might use tin snips, a Sawzall, a jigsaw, or a cutting torch. On wood decks you might use a Sawzall, jigsaw, or keyhole saw.

When working on a concrete roof deck that will be poured in place, set cans on the deck forms. Cans are short pieces of pipe or wooden boxes that preserve roof drain openings as the concrete is poured (see *Figure 20*). Be sure to install the cans before the concrete is poured so you won't have to drill these openings later.

6.2.0 Attaching the Drain Body

Insert the drain body through the opening you cut and secure it with a deck clamp (see *Figure 21*). For installations over rigid insulation or in cases where the opening is oversized, install a drain receiver before putting the drain body in position. The drain receiver provides a wide flange to support the drain.

6.3.0 Connecting to the Piping System

You can make the joint between the drain body and the storm water piping system in several ways. If the piping material is cast iron, the joint may be caulked, no-hub, compression, or threaded. The procedures for making these joints are identical to those used in joining cast-iron pipe (see *Figure 22*). You can join plastic roof drain bodies to the piping system with solvent welding or threaded joints.

Cut the Right Size Hole

All drains have a certain size hole into which they will fit. If you cut the hole too large, you may need to repair the roof deck or buy a bigger drain, which will cost more in both time and money.

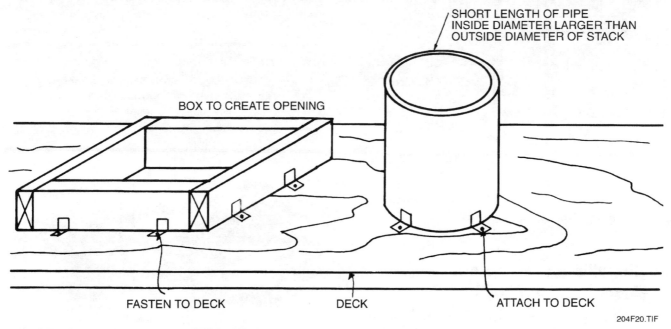

Figure 20 ◆ Cans preserve roof drain openings.

6.4.0 Installing Expansion Joints

To deal with potential movement between the roof deck and the rest of the structure, install an expansion joint (see *Figure 23*). This joint is installed in the vertical pipe between the drain body and the horizontal piping.

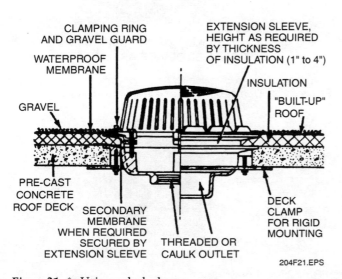

Figure 21 ◆ Using a deck clamp.

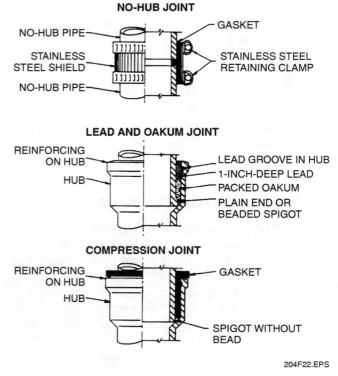

NO-HUB JOINT

LEAD AND OAKUM JOINT

COMPRESSION JOINT

Figure 22 ◆ Joints for cast-iron roof drains.

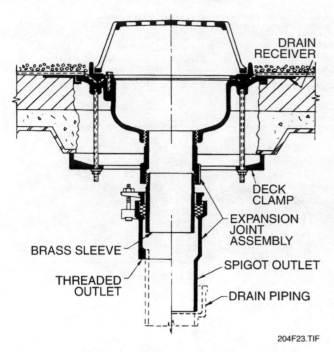

Figure 23 ◆ Expansion joint installation.

6.5.0 Installing a Waterproof Membrane or Flashing

Specifications may require installation of a waterproof membrane or flashing. A roofer may do this or it may be your responsibility. If it is, follow these steps:

Step 1   Place the flashing or waterproof membrane around the drain after installing the roof drain.

Step 2   Make an opening the size of the drain.

Step 3   Secure the membrane or flashing with the flashing clamp.

Depending on the roof design, you may place the waterproof membrane either above or below the insulation.

6.6.0 Checking the Roof Drains

You are responsible for ensuring that the roofers' work, as it relates to your job, is completed satisfactorily. Once the roof is completed, check the roof drains to make sure that they are not clogged with roofing material. Also, make sure that the roof drain domes are properly secured. Make sure that adjustable flow roof drains are at the correct setting (see *Figure 24*).

A problem you'll encounter frequently is maintaining the correct elevation on adjustable roof drains. The compression joint used for some roof drains is easy to install (see *Figure 25*). However, adjustable drains and compression fittings are easily displaced if someone steps on the drain. If this happens, you will have to reposition the drain and have part of the roof replaced—a major, and costly, repair.

CLAMP RING WITH STATIONARY VANES

ADJUSTABLE VANE COLLAR

COMPLETE FLO-SET CONTROL DEVICE ASSEMBLY

204F24.EPS

Figure 24 ◆ Domes of adjustable flow roof drains.

204F25.TIF

Figure 25 ◆ Compression joint.

Review Questions

Sections 5.0.0–6.6.0

1. The best time to set the elevation of a floor drain in a concrete floor is _____.
    a. after the concrete floor is poured
    b. after finished flooring is installed
    c. before the concrete floor is poured
    d. before the building foundation is finished

2. The elevation of a floor drain should be measured with a(n) _____.
    a. T-square and a steel ruler
    b. surveyor's level or transit
    c. architect's ruler
    d. folding rule

3. The best way to support a drain at the proper elevation in a concrete floor before the floor is poured is with _____.
    a. tamped fill
    b. loose gravel
    c. wooden blocks
    d. concrete blocks or bricks

4. A waterproof membrane is needed in a floor drain that is not located in a _____.
    a. bathroom
    b. basement
    c. kitchen
    d. tub or shower

5. The best way to create the opening for roof drains that will be installed in a concrete roof deck is to _____.
    a. drill each opening after the roof deck has cured
    b. frame the openings in pipe or wood before the roof deck is poured
    c. ask the concrete crew to create the openings
    d. mark where the roof drains are needed with cans after the deck is poured

Summary

Roof, floor, and area drains are similar in design and installation. In each case, refer to specifications or plumbing drawings for information on the type of drains to install and where to install them. Always consult local codes, and in areas that require high levels of sanitation, check with the local health department. Because drain installation work will overlap the work of roofers, concrete workers, and other trades, talk with them and coordinate your jobs.

The precise location of drains is vital because poorly located drains will not function properly. Once you've positioned the drains at the proper elevation, protect them from movement, damage, and clogging during the construction process.

You will use conventional pipe-joining techniques to connect roof, area, and floor drain piping. You'll use many different sizes of pipes and a variety of pipe materials. Consult the local codes and job specifications.

Trade Terms Introduced in This Module

Can: A short length of pipe or wooden box attached to roof decks before concrete is poured to preserve roof drain openings.

Can wash drain: A can that is placed upside down over a drain to allow water to squirt upward, washing the inside of the container.

Clamping ring: A clamp that secures subsurface flashing or waterproof membrane to the drain body.

Deck clamp: A ring that fastens to a drain body with several bolts.

Dome: A ventilated drain cover that prevents debris from entering the piping system and helps reduce the chance of pipe clogs. Also called *grate* or *strainer*.

Drain body: The basic element of all drains, the drain body funnels water into the piping system.

Drain receiver: A type of flange that distributes the roof drain's weight over a large area. It also supports a drain in an oversized or off-center roof opening.

Expansion joint: A joint or gap between adjacent parts of a building that permits them to move as a result of temperature changes or other conditions without damage to the structure.

Extension: A part that extends a fixture, such as a drain, to make it flush with the finished floor or roof.

Floor sink: A type of floor drain that is coated with a porcelain or easy-to-clean finish. Most often used in areas requiring a high level of sanitation.

Flushing floor drain: A floor drain that incorporates an inlet in the trap to allow connection of a water supply pipe to flush the trap.

Grate: A ventilated drain cover that prevents debris from entering the piping system and helps reduce the chance of pipe clogs. Also called a *dome*.

Sediment bucket: A removable device inside a drain body that traps small solids that pass through the grate, or dome, to keep the solids out of the piping. Also called a *sediment trap*.

Trap primer: A primer that periodically adds a small amount of water to the trap. Often used in traps where limited use is expected.

Trap primer valve: A valve that allows a small amount of water into a trap when the valve is operated.

Answers to Review Questions

Sections 2.0.0–4.3.0
1. d
2. c
3. a
4. b
5. a

Sections 5.0.0–6.6.0
1. c
2. b
3. d
4. b
5. b

ACKNOWLEDGMENTS

Figure Credits

Josam Manufacturing   204F01, 204F02, 204F03, 204F04, 204F05,
204F06, 204F07, 204F08, 204F09, 204F10,
204F15, 204F16, 204F17

Zurn Industries, Inc.   204F11, 204F12, 204F13, 204F14, 204SA01,
204SA02, 204F19, 204SA03, 204F21

NCCER CRAFT TRAINING USER UPDATES

The NCCER makes every effort to keep these textbooks up-to-date and free of technical errors. We appreciate your help in this process. If you have an idea for improving this textbook, or if you find an error, a typographical mistake, or an inaccuracy in the NCCER's Craft Training textbooks, please write us, using this form or a photocopy. Be sure to include the exact module number, page number, a detailed description, and the correction, if applicable. Your input will be brought to the attention of the Technical Review Committee. Thank you for your assistance.

Instructors – If you found that additional materials were necessary in order to teach this module effectively, please let us know so that we may include them in the Equipment and Materials list in the Instructor's Guide.

Write: Curriculum Revision and Development Department
National Center for Construction Education and Research
P.O. Box 141104, Gainesville, FL 32614-1104

Fax: 352-334-0932

E-mail: curriculum@nccer.org

Craft _____ Module Name _____

Copyright Date _____ Module Number _____ Page Number(s) _____

Description _____

_____

_____

_____

(Optional) Correction _____

_____

_____

(Optional) Your Name and Address _____

_____

_____

Types of Valves

COURSE MAP

This course map shows all of the modules in the second level of the Plumbing curriculum. The suggested training order begins at the bottom and proceeds up. Skill levels increase as you advance on the course map. The local Training Program Sponsor may adjust the training order.

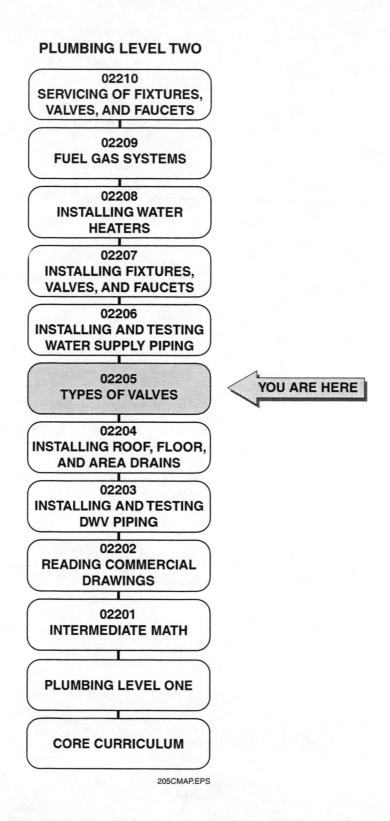

PLUMBING LEVEL TWO

02210
SERVICING OF FIXTURES, VALVES, AND FAUCETS

02209
FUEL GAS SYSTEMS

02208
INSTALLING WATER HEATERS

02207
INSTALLING FIXTURES, VALVES, AND FAUCETS

02206
INSTALLING AND TESTING WATER SUPPLY PIPING

02205
TYPES OF VALVES   ◁ YOU ARE HERE

02204
INSTALLING ROOF, FLOOR, AND AREA DRAINS

02203
INSTALLING AND TESTING DWV PIPING

02202
READING COMMERCIAL DRAWINGS

02201
INTERMEDIATE MATH

PLUMBING LEVEL ONE

CORE CURRICULUM

205CMAP.EPS

Figures

Tables

Types of Valves

Objectives

When you have completed this module, you will be able to do the following:

1. Identify the basic types of valves.
2. Describe the differences in pressure ratings for valves.
3. Demonstrate the ability to service various types of valves.

Prerequisites

Before you begin this module, it is recommended that you successfully complete the following modules: Core Curriculum; Plumbing Level One; Plumbing Level Two, Modules 02201 through 02204.

Required Trainee Materials

1. Paper and sharpened pencil
2. Appropriate Personal Protective Equipment

1.0.0 ◆ INTRODUCTION

Valves regulate flow. They may provide on/off service, act as a throttling device, or prevent flow reversal through a line. The most common types of valves are **gate**, **globe**, **angle**, **ball**, **butterfly**, and **check valves**.

2.0.0 ◆ HOW VALVES OPERATE

Valves use one or more of the following four methods to control flow through a piping system:

- Move a disc or plug into or against a passageway.
- Slide a flat cylindrical or spherical surface across a passageway.
- Rotate a disc or ellipse around a shaft extending across the diameter of a pipe.
- Move a flexible material into the passageway.

2.1.0 Valve Terminology

It is important use the correct valve terms. Study the following:

Trim – The parts of a valve that receive the most wear and tear and, consequently, are replaceable. Trim includes the stem, disc, seat ring, disc holder (or guide), wedge, and bushings.

Straight-through flow – Describes an unrestricted flow. The element that closes the valve retracts until the passage is clear.

Full flow – Designates the relative flow capacities of various valves.

Throttled flow – A valve designed to partially control or throttle the volume of liquid. Not all types of valves are suitable for throttled flow. If you choose the wrong valve, it will be difficult to control flow and the valve will wear out very quickly.

3.0.0 ◆ TYPES OF VALVES

The six common types of valves are as follows:

- Gate valves
- Globe valves
- Angle valves
- Ball valves
- Butterfly valves
- Check valves

Purpose of Valves

Although valves operate differently, they all serve the same purpose—to control the flow of liquids and gases.

3.1.0 Gate Valves

A gate valve controls flow with a gate, or disc, that slides in machined grooves at right angles to the flow. The action of the threaded stem on the control handle moves the gate.

Gate valves are best suited for main supply lines and pump lines. With the gate fully opened, the valve provides an unobstructed passageway. Install gate valves on lines containing steam, water, gas, oil, and air.

Gate valves contain either a solid wedge or a split wedge (see *Figures 1* and *2*). Use **solid-wedge valves** for steam, hot water, and other services where shockwave is a factor. The wedge is precision-machined at an angle that matches the body seat. This angle conforms to the taper of the body seat to ensure a tight seal.

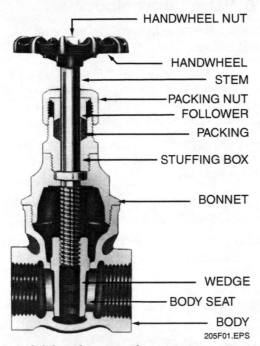

Figure 1 ◆ Solid-wedge gate valve.

Use **split-wedge valves** on lines where you need a more positive closure—for example, on low-pressure lines and cold water lines. Also use split-wedge valves on fluid lines such as gasoline, benzene, kerosene, and solvents, all of which are light, volatile, and difficult to handle. The wedge's ball-and-socket joint helps align both faces against the body seat. When the wedge lowers com-

pletely, the downward motion of the stem forces the wedged halves outward against the body seats, forming a tight closure. When the valve opens, it relieves pressure on the split wedge. The split wedge is then removed by the upward motion of the stem from contact with the seats.

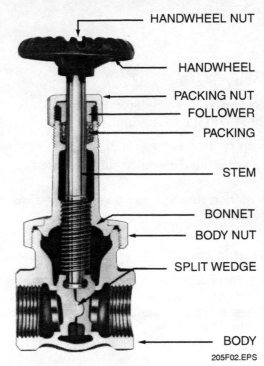

Figure 2 ◆ Split-wedge gate valve.

The split-wedge design retains better closure than the solid-wedge design in cases of sudden temperature changes or flexing in the line.

3.1.1 Advantages and Disadvantages of Gate Valves

The advantages of gate valves are as follows:

- They provide maximum unobstructed full flow.
- They have minimum pressure drop.
- They can be completely cleared from the line of flow.

You can avoid the disadvantages of gate valves with proper precautions and care. Following are some disadvantages and steps you can take to prevent problems.

- Material trapped between the wedge and the seat can damage the valve so that it won't close properly. To avoid this problem, close the valve slowly just before the wedge meets the seat. This creates a greater flow between the wedge and the seat that washes away any trapped solid materials.
- Liquids can remain trapped in the valve after you drain the lines. Liquid that freezes inside the valve expands and can destroy it. To avoid this, drain the valves as well as the lines.
- To avoid other gate valve problems, open and close the valve slowly and completely. Because approximately 80 percent of full flow occurs in the first 20 percent of opening, a sudden release of pressure can create a shock force commonly called **water hammer**.

3.2.0 Globe Valves

A globe valve controls flow by moving a circular disc against a metal seat that surrounds the flow opening (see *Figure 3*). The screw action of the turning handle forces the disc onto the seat or draws it away.

Use globe valves for general service on steam, water, gas, and oil lines that require frequent operation and close flow control. A partition inside the valve body closes off the valve's inlet side from the outlet side, except for a circular opening called the *valve seat*. The upper side of the valve seat is ground smooth to ensure a complete seal when the valve is closed. To close the valve, turn the handwheel clockwise until the valve stem firmly seats the washer, or disc, between the valve stem and the valve seat. This stops the flow of gas or liquid.

Threads on the stem screw into corresponding threads in the valve's upper housing. The top of the housing is hollowed out so that it can hold a **packing** material. You can replace this packing if a leak occurs between the packing nut and the valve stem.

3.2.1 Advantages and Disadvantages of Globe Valves

The major advantages of globe valves are as follows:

- Their critical parts (washer, seat, and packing) are replaceable.
- They accurately control water flow.
- They are durable and won't wear out easily despite repeated rapid use.

Because globe valves are reliable and easy to repair, plumbers frequently install them in a building's water supply lines. They are used, for example, to cut off the hot or cold water supply to a bathroom while a fixture is repaired.

In spite of their popularity, globe valves have two major disadvantages:

- They partially obstruct flow even when they are fully opened.
- You cannot drain them completely, so any liquid remaining in the valves could freeze, expand, and damage the valve.

3.3.0 Angle Valves

The angle valve is similar to the globe valve but can serve as both a valve and a 90-degree elbow (see *Figure 4*). Because flow changes direction only twice through an angle valve, the valve is less resistant to flow than the globe valve in which flow must change direction three times. Angle valves come with conventional, plug-type, or composition discs.

3.4.0 Ball Valves

The ball valve controls the flow of gases and liquids (see *Figure 5*). Install it in piping systems where you want quick shutoff. Also use it for in-line maintenance or in lines in which liquids and gases are mixed. A handle on the outside of the valve body rotates the valve's ball part into the opened or closed position. These valves provide positive, quick flow control on piping systems.

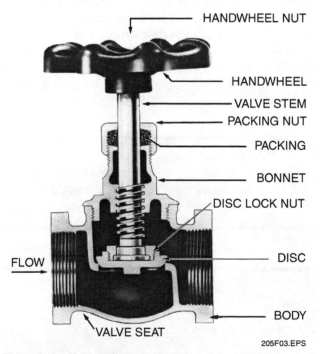

HANDWHEEL NUT
HANDWHEEL
VALVE STEM
PACKING NUT
PACKING
BONNET
DISC LOCK NUT
FLOW
DISC
BODY
VALVE SEAT
205F03.EPS

Figure 3 ◆ Globe valve.

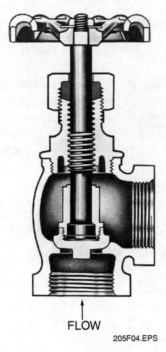

FLOW

205F04.EPS

Figure 4 ◆ Angle valve.

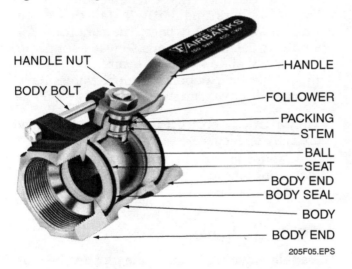

HANDLE NUT
BODY BOLT
HANDLE
FOLLOWER
PACKING
STEM
BALL
SEAT
BODY END
BODY SEAL
BODY
BODY END

205F05.EPS

Figure 5 ◆ Ball valve.

3.5.0 Butterfly Valves

A butterfly valve is built on the principle of a pipe damper (see *Figure 6*). A disc about the same diameter as the inside diameter of the pipe controls the flow. The disc rotates either vertically or horizontally and seals on a seat machined on the inside diameter of the pipe. For throttling service, secure the disc in place with handle-locking devices.

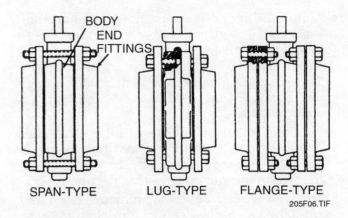

SPAN-TYPE   LUG-TYPE   FLANGE-TYPE

205F06.TIF

Figure 6 ◆ Butterfly valves.

Butterfly valves, especially the larger sizes, offer many advantages over the valves already discussed. They are lighter, take up less space, cost less, have fewer moving parts, and contain no pockets to trap fluids.

Use butterfly valves either fully open or fully closed for throttling service and for frequent operation. In general, they do a good job of handling **slurries** (liquids that contain large amounts of suspended solids). In addition, they offer a positive shutoff for gases and liquids.

3.6.0 Check Valves

A check valve prevents flow reversal in a piping system. Pressure in the line keeps the valve opened. Flow reversal and the weight of the disc mechanism will automatically close the valve.

The types of check valves available include the **ball-check valve**, the **swing-check valve**, and the **lift-check valve**. The ball-check valve allows one-way flow in water supply or drainage lines. You can use it in lines with extremely low back pressure (see *Figure 7*).

 CAUTION

When operating quarter- or half-turn valves, you should open or close them slowly to avoid water hammer.

ON THE
LEVEL

Positive Shutoff

When valves are described as offering a positive shutoff, it means that they positively will not leak when shut off.

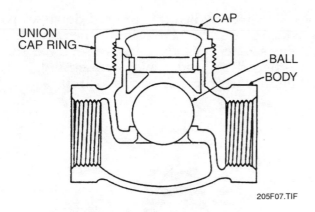

Figure 7 ◆ Ball-check valve.

The swing-check valve features a low flow resistance, which makes it suited for lines containing liquids or gases with low to moderate pressures (see *Figure 8*). Depending on the manufacturer, the swing-check valve comes in four different types: bronze mounted, all iron, rubber-faced disc, and **lever-and-weight**.

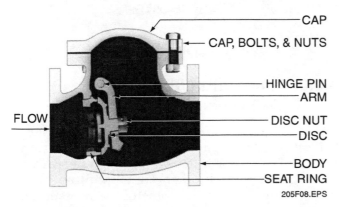

Figure 8 ◆ Swing-check valve.

Plumbers use the bronze-mounted swing-check valve most often. Because it is soft and pliable, the bronze facing offers a more positive sealing effect than facings made from harder materials. Use the all-iron swing-check valve for services where bronze might erode. Any fluid or gas that contains corrosive chemicals will eventually erode brass.

Use a rubber-faced swing-check valve when noise in the check valve disc seating is a problem. This non-slam disc muffles the closing sound of the valve. It also helps protect the seat when high head pressures sharply reverse the flow.

Use the lever-and-weight swing-check valve when pulsation occurs in the lines. You can install these valves in either a horizontal or a vertical position and adjust them to prevent chatter from turbulence or pulsation in the line. The disc also adjusts to open at different amounts of pressure.

Use the lift-check valve for gas, water, steam, or air, and for lines where frequent fluctuations in flow occur (see *Figure 9*). These valves come in horizontal and vertical styles. The integral construction of the horizontal type is similar to the globe valve. The vertical type allows a straight- through flow.

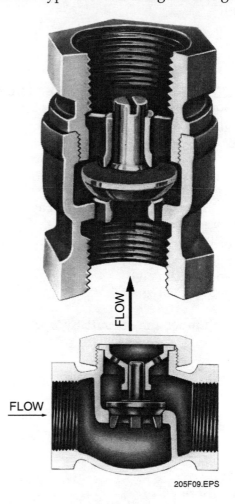

Figure 9 ◆ Vertical and horizontal lift-check valves.

 DID YOU KNOW?

Valve Component Terms

Terms for valve components vary widely depending on the manufacturer, region, and the size or type of valve. For example, followers are used in certain types of valves to put pressure on the packing when the packing nut is turned. The follower is also called a packing gland or simply a gland. Flanges that incorporate more than one nut to tighten followers are also called follower flanges, packing flanges, stem flanges, and gland flanges. Take the time to learn the valve terms used in your area.

Review Questions

Sections 2.0.0–3.6.0

1. The part of a valve that receives the most wear and tear is the _____.
   - a. trim
   - b. seat
   - c. disc
   - d. packing

2. A(n) _____ is most commonly used to control bathroom water supply lines.
   - a. angle valve
   - b. ball valve
   - c. check valve
   - d. globe valve

3. A _____ is best to use on lines carrying hot water or steam.
   - a. split-wedge gate valve
   - b. solid-wedge gate valve
   - c. swing-check valve
   - d. ball-check valve

4. Water hammer may result by opening gate valves too _____.
   - a. quickly
   - b. slowly
   - c. far
   - d. little

5. A _____ is best when a throttled flow is needed in a piping system.
   - a. ball valve
   - b. ball-check valve
   - c. butterfly valve
   - d. lift-check valve

4.0.0 ◆ SPECIAL VALVES

Special valves are designed for one purpose only. Special valve applications include **pressure regulator valves**; **temperature/pressure (T/P) relief**, or safety, valves; **supply stop valves**; **float-controlled valves/ball cocks**; **flush valves**; and **flushometers**.

4.1.0 Pressure Regulator Valves

The pressure regulator valve reduces water pressure in a building (see *Figure 10*). Pressure changes in the system activate the valve. As the pressure changes, a spring located in the valve's dome acts on a diaphragm to move the valve up or down.

The valve opens when it is pressed down away from the valve seat and remains opened until the building's water pressure reaches a set level. The valve then closes and remains closed until the building's water pressure begins to drop.

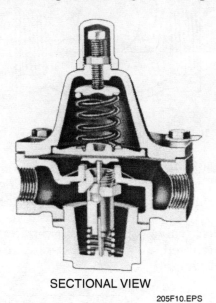

SECTIONAL VIEW

205F10.EPS

Figure 10 ◆ Pressure regulator valve.

4.2.0 Temperature/Pressure (T/P) Valves

The terms *temperature/pressure (T/P) relief valve* and *T/P safety valve* are used interchangeably. Spring-loaded T/P safety valves and T/P relief valves look similar to each other. They both limit fluid pressure by discharging some of the pressurized liquid or gas.

 WARNING!
Do not use a relief valve in a system that requires instantaneous release of large volumes of steam, compressed air, or other gases. A regular relief valve could seize up or fail to function and an explosion might result.

Use steam safety valves in lines with gases that carry air and steam (see *Figure 11*). Their design includes a **huddling chamber** that harnesses the expansion forces of these gases to quickly open (pop) or close the valve. The difference between the opening and closing pressures is called **blowdown**. Blowdown limitations for steam safety valves are stated in the **ASME Power Boiler Vessel Code**.

205F11.EPS

Figure 11 ◆ Steam safety valve.

205F12.EPS

Figure 12 ◆ Relief valve.

Use relief valves for liquid service (see *Figure 12*). Ordinarily, relief valves do not have an accentuating huddling chamber or a regulator ring to vary or adjust blowdown. They therefore operate with a relatively lazy motion. As pressure increases they slowly open, and as pressure decreases they slowly close. In vessels or systems that don't require instantaneous release of large volumes, this valve provides sufficient protection. It is also appropriate where sufficient leeway is provided between the design pressure and the operating pressure in the system.

4.3.0 Supply Stop Valves

Use supply stop valves to disconnect the hot or cold water supply to individual fixtures such as water closets, lavatories, and sinks. These valves, also called *supply valves*, make it easy to control the water connection at an individual fixture, making repair or replacement work easier as well. They are chrome-plated for an attractive appearance, and they come in either straight or right-angle designs (see *Figure 13*).

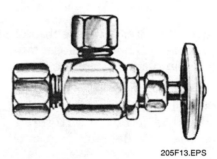

205F13.EPS

Figure 13 ◆ Supply stop valve.

4.4.0 Float-Controlled Valves/Ball Cocks

Float-controlled valves, installed in water closet flush tanks, maintain a constant water level in the tank (see *Figure 14*). Their design may vary, but all use the water level in the tank to control flow. When the water level drops, the valve lifts up and away from the seat as the arm lowers. The float arm then rises as the water level in the tank rises. This forces the valve down against the seat, which closes the valve and stops the flow of water.

ON THE
· LEVEL ·

Repair or Replace?

The cost of materials and labor make repairing many valves uneconomical. Installing a new valve may make more sense. Note, however, that some valves such as T/P valves should always be replaced.

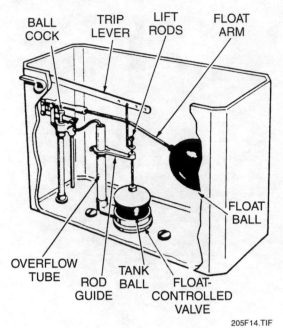

Figure 14 ◆ Float-controlled valve.

To install the float valve assembly, follow these steps:

Step 1 Lower the threaded base through the bottom hole in the water closet tank with the gasket in place against the flange.

Step 2 Place the washer and nut on the base of the assembly on the outside of the tank and tighten. Do not overtighten the nut because the toilet tank may crack.

Step 3 Place the riser and coupling nut on the base of the float valve and hand tighten.

Step 4 Shape the riser to align with the shutoff valve, remove the riser, and cut it to fit.

Step 5 Reattach the riser to the tank and the shutoff valve and turn on the water supply.

Step 6 Adjust the float ball in the tank to achieve the water level shown on the tank.

Step 7 Adjust the float ball with the adjusting screw. Do not bend the float arm because the rod may work around a half-turn (180 degrees) and cause the tank to overflow.

4.5.0 Flush Valves

Flush valves, also installed in water closet flush tanks, control the flow of water from the tank into the bowl (see *Figure 15*). When you depress the tank lever, the valve lifts above the tank outlet and floats there. This allows the water in the tank to flow rapidly into the bowl. When the water level in the tank drops to the point where the flush

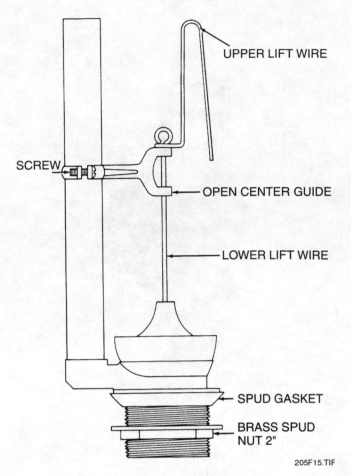

Figure 15 ◆ Flush valve.

valve no longer floats, the valve reseats on the tank drain and the tank—regulated by the float-controlled valve—refills.

4.6.0 Flushometers

Flushometers, used for water closets or urinal fittings, require no storage tank because the water flows, under pressure, directly into the fixture. The scouring action of the water created when the flushometer opens generally cleans more effectively than the gravity flow from a storage tank. Flushometers are popular in commercial installations. Plumbers can connect them directly to the water supply pipe, and they can flush repeatedly in a short space of time without waiting for a storage tank to refill.

Flushometers are available in diaphragm or piston types. Both work the same way. In the diaphragm type, water flow stops when water pressure in the upper chamber forces the diaphragm against the valve seat (see *Figure 16*). Motion of the handle in any direction pushes the plunger against the auxiliary valve. Even the slightest tilt will cause water leakage from the upper chamber around the diaphragm.

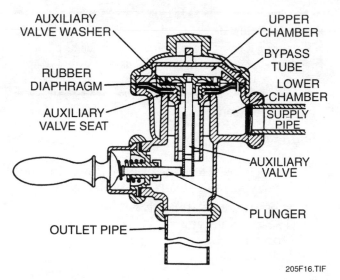

Figure 16 ◆ Diaphragm type flushometer.

When pressure lessens in the upper chamber, the diaphragm rises, allowing water to flow into the fixture. While the fixture is in flushing mode, a small amount of water flows through the bypass. As the water fills the upper chamber, the diaphragm reseats against the valve seat. This action shuts off the flow of water to the fixture.

 DID YOU KNOW?
Variety of Valves

Valves come in many different sizes. Very small ones are used in automobile and bicycle tires. Huge valves control the flow of water in canal locks. No matter the size, all valves have two basic parts—an enclosed body containing a hole through which fluids or gases pass, and a disc that opens and closes the hole.

5.0.0 ◆ MATERIALS

Valves are made from practically every machinable material. The most common materials used are iron, brass, carbon steel, bronze, aluminum, alloy steels, and polyvinyl chloride (PVC). Circumstances dictate which material to use in a particular application. Generally, more than one material will satisfy the requirements of a given situation.

For example, valves used in water service are made chiefly from bronze, brass, malleable iron, cast iron, copper, forged steel, stainless steel, thermoset plastic, and thermoplastic. Frequently, different materials make up the various parts of a single valve. The valve body may be formed from cast iron, the valve plug from cast bronze, the O-ring seal from rubber, and the compression washer from Teflon®.

Air and water service valves are made of bronze or iron with bronze trim. Valves used for low-pressure steam applications are made of iron. Valves that regulate noncorrosive products are made of steel.

Although expensive, stainless steel is required for valves that regulate the flow of corrosive substances. Stainless steel is also used when it is necessary to prevent contamination of fluids flowing through a line. Most valves used in cryogenic (very low temperature) applications are made from stainless steel because it does not get as brittle as iron or carbon steel at low temperatures. Some chemicals—for example, chlorine and sulfuric acid—require specialized valve bodies. These are made from plastic, rubber, ceramics, or special alloys.

 DID YOU KNOW?

Brass is an alloy (mixture) of copper and zinc. The first use of brass was around 700 B.C.E. by the people of the Middle East. During the 18th and 19th centuries, hundreds of different copper alloys were developed. Most of them were brasses. Today, brass is the most widely used copper alloy.

5.1.0 Selecting Valves

Many factors contribute to the proper selection of valves. What is the valve's purpose? If the valve will provide on/off service, select a gate valve, butterfly valve, or ball valve. If the valve will throttle and regulate the flow, select a globe valve, butterfly valve, or ball valve. Once you have determined the valve's purpose, analyze the material in the system. Some of the factors to consider are the following:

- Is the material a liquid or a gas?
- How freely can the material move through the system?
- Is the material abrasive?
- Is the material corrosive?
- What is the temperature of the material?
- What is the pressure in the system?
- To what extent must the valve be leak-tight?
- What is the maximum pressure drop the system can tolerate?

 WARNING!
Be sure that the valve you use is suited for the intended system. You cannot use bronze or cast-iron materials in corrosive systems.

5.2.0 Sizing

Sizes of valves vary according to diameter, length, width, and height. Standard diameters range from ⅛-inch to ½-inch, with ¾-inch to ½-inch commonly used in residential installations and 1½-inch to 2-inch commonly used in commercial installations.

It is often difficult to determine which size valve is required for a given application. The general rule of thumb is to use a valve with the same connection size as the pipeline in which it is installed. The problem with this method is that sometimes the capacity of the valve is overlooked and the system may operate inconsistently. Base valve sizing on valve capacity and known performance values.

6.0.0 ◆ VALVE RATINGS

A code marked on the valve body indicates which gases or liquids it is designed to carry. The code is as follows:

- O = oil
- W = water
- S = steam
- G = gas
- L = liquid other than water
- SWP = steam working pressure

For example, a valve marked *125 SWP with 200 WOG* operates safely at 125 psi (pounds per square inch) of saturated steam and 200 psi of cold water, oil, or gas.

Review Questions

Sections 4.0.0–6.0.0

1. A _____ is best to control blowdown in lines that carry air and steam in a boiler system.
   a. pressure regulator valve
   b. T/P relief valve
   c. stop valve
   d. ball valve

2. A _____ delivers water from a water closet storage tank until the valve reseats into the tank drain.
   a. supply stop valve
   b. flushometer
   c. T/P relief valve
   d. float-controlled valve

3. A _____ delivers water under pressure without a storage tank to a water closet or urinal.
   a. globe valve
   b. flushometer
   c. float-controlled valve
   d. pressure regulator valve

4. Use _____ to disconnect the hot or cold water supply to individual fixtures such as water closets, lavatories, and sinks.
   a. gate valves
   b. globe valves
   c. supply stop valves
   d. ball check valves

5. A valve that is marked *100 SWP with 200 WOG* operates safely at _____.
   a. 100 psi of saturated steam with 200 psi of cold water, oil, or gas
   b. 100 psi of steam and 200 psi of any fluid except water
   c. 100 psi of steam made from salt water and 200 psi of water or gas
   d. 100 psi of saturated, steamed cold water with 200 psi of natural gas

ON THE LEVEL

Properly Rated Valves

Always check valve ratings to make sure you have the properly rated valve for a particular system.

7.0.0 ◆ TYPES OF STEMS

The four basic types of stems used on valves are as follows:

- **Rising stem**
- **Nonrising stem**
- **Outside screw and yoke stem**
- **Sliding stem**

7.1.0 Rising Stem

In a rising stem unit, both the handwheel and the stem rise (see *Figure 17*). The height of the stem gives an approximate indication of how far the valve is open. However, you can install valves with rising stems only in areas with sufficient headroom. The stem itself comes into contact with the fluid in the line.

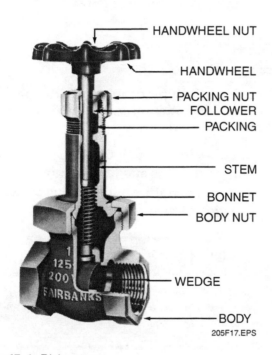

Figure 17 ◆ Rising stem.

7.2.0 Nonrising Stem

As the name implies, neither the handwheel nor the stem rises when the valve opens. Nonrising stems are suitable for areas where space is limited (see *Figure 18*). Because only a spindle inside the valve body turns when the handwheel turns, stem wear is kept to a minimum. This type of stem also makes contact with the fluid in the line.

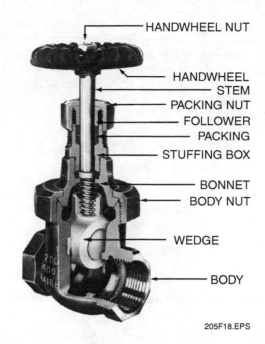

Figure 18 ◆ Nonrising stem.

7.3.0 Outside Screw and Yoke Stem

Often abbreviated OS&Y, the outside screw and yoke stem is suitable for use with corrosive fluids because it does not come in contact with the fluid in the line (see *Figure 19*). As the handwheel turns, the stem moves up through it. The height of the stem gives an approximate indication of how far the valve is open.

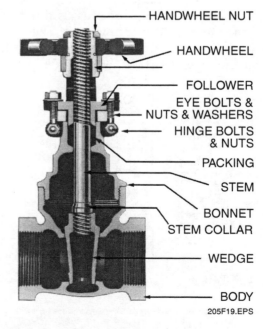

Figure 19 ◆ Outside screw and yoke stem.

7.4.0 Sliding Stem

Use the sliding stem when quick opening and closing is required (see *Figure 20*). For example, the sliding stem could be used when filling large tanks or vats.

205F20.EPS

Figure 20 ◆ Sliding stem.

7.5.0 Packing

Achieving a good seal at the stem of a valve is somewhat difficult because the stem moves.

Packing seals the stem and prevents fluid from leaking up through it. Packing also retains the pressure of the fluid in the valve. The packing material fills the stuffing box, a space between the valve stem and **bonnet**. A follower presses the packing against the stem in the stuffing box. The stuffing box requires occasional tightening, especially if the valve has not been used for a while (see *Figure 21*).

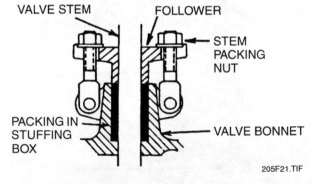

Figure 21 ◆ Packing.

205F21.TIF

7.5.1 Types of Packing

There are many different kinds of packing. The more common types are as follows:

- Solid
- Braided
- Granulated fibers

Choosing the type of material for packing involves consideration of many factors. Some of these are as follows:

- Material flowing through the line
- Operating pressures and temperatures
- Minimum temperature of the piping system
- Composition of the valve stem

CAUTION

Be sure that the packing is suited for the material flowing through the line and the operating pressures and temperatures. Improper packing will fail and leaks will occur.

8.0.0 ◆ BONNETS

The valve bonnet is a cover that guides and encloses the valve stem. The three basic types of bonnets are as follows:

- **Screwed bonnet**
- **Union bonnet**
- **Bolted bonnet**

ON THE LEVEL

Stem Packing Nut

When new, the stem packing nut on smaller valves requires tightening or it will leak.

ON THE

LEVEL

Bonnets and Gaskets

Some bonnets do not need a gasket. They have a machined seating surface where they thread to the valve.

8.1.0 Screwed Bonnet

The screwed bonnet is the simplest and least expensive bonnet design (see *Figure 22*). A two-piece configuration, it is frequently used in low-pressure applications and where periodic disassembly of the valve is not required. Screwed bonnets are often used on bronze gate, globe, and angle valves.

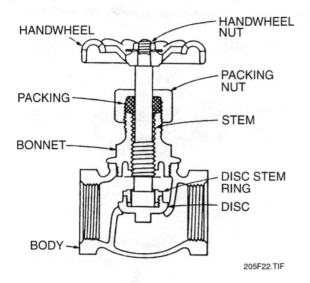

Figure 22 ◆ Screwed bonnet.

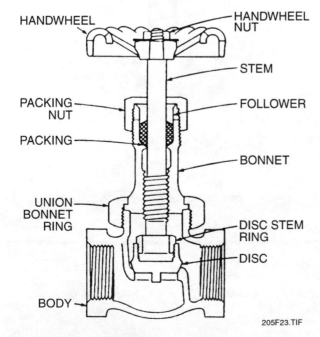

Figure 23 ◆ Union bonnet.

8.2.0 Union Bonnet

The three-piece union bonnet is best suited for those applications where frequent disassembly of the valve is required (see *Figure 23*). A separate union ring holds the valve bonnet to the valve body, giving the body added strength against internal valve pressures. The union bonnet works well with the smaller sizes of valves, but is impractical for the larger sizes.

WARNING!

Do not use a union bonnet in high-pressure applications.

8.3.0 Bolted Bonnet

The bolted bonnet joint is used frequently for larger sizes of valves or for high-pressure applications (see *Figure 24*). A series of small-diameter bolts fastens this type of bonnet to the body. This allows a uniform sealing pressure and maintenance with small wrenches.

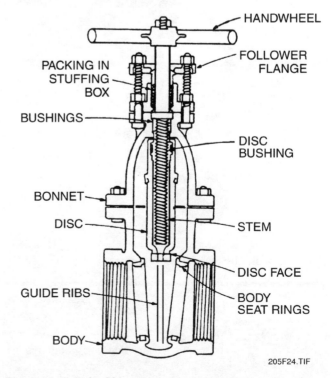

Figure 24 ◆ Bolted bonnet.

9.0.0 ◆ END CONNECTIONS

Valves are manufactured with four basic types of end connections:

- **Internal threads**
- **Solder ends**
- **Flanged ends**
- **External threads**

Most metal-bodied valves are manufactured with internal threads that allow for easy hookup to threaded pipe (see *Figure 25*).

Use solder end valves to hook up to copper and brass pipes that have no internal or external threads. You can purchase these valves in bronze or bronze mounted and copper (see *Figures 26* and *27*).

In areas where ease of changing valves is necessary, use valves with flanged ends (see *Figure 28*). It's quicker and more economical than resoldering. These valves are generally used on large pipes (4 inches and larger) and in commercial and industrial installations.

The external thread design is frequently used on plastic valves. These valves require a nut in order to make a compression joint with the pipe (see *Figure 29*).

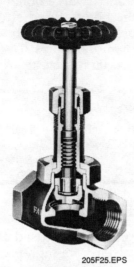

205F25.EPS

Figure 25 ◆ Threaded connection.

205F27.EPS

Figure 27 ◆ Copper valve with solder end.

205F26.EPS

Figure 26 ◆ Solder end valve.

COURTESY OF CRANE COMPANY

205F28.EPS

Figure 28 ◆ Flanged end.

ON THE LEVEL

Solder End

The most common end connection in plumbing systems is the solder end. This is because of the wide and varied use of copper tube.

Figure 29 ◆ Plastic valve.

205F29.EPS

CAUTION

Always check end connections for defects or dirt and grit before installation.

DID YOU KNOW?

During the Industrial Revolution, industries became major users of valves. These industries produced items such as textiles, chemicals, and processed food. With the birth of the petroleum industry came increased demand for high-performance valves. These valves had to handle the great pressure of oil and gas flowing from deep wells to the surface. As demand for higher performance valves grew, manufacturers responded with improvements in engineering, materials, and design.

10.0.0 ◆ REPAIRING VALVES

You now have a basic understanding of how valves operate. To service or repair valves, you need to identify the problem, determine what caused it, and efficiently make the repair. Experience will help you develop the knowledge and skills to select and use the right tools and efficiently repair or replace parts.

10.1.0 General Safety Guidelines

The severity of the problems you'll encounter when responding to a service call will vary greatly.

For example, a small leak from a faucet spout into a kitchen sink, while wasteful and annoying, is not an emergency. On the other hand, a valve leaking a lot of water above a suspended ceiling, is. In an emergency like this, you must stop the flow of water, minimize the potential damage, and make the area safe before you continue with the repairs.

Some general guidelines for repair work follow. As you progress in your career, add your experiences to these guidelines.

WARNING!

Always take safety precautions when repairing plumbing components. Do not assume that the electricity has been turned off. Always check for yourself.

- Wear rubber-soled shoes or boots for protection from slipping and electric shock.
- Turn off electrical circuits.
- Shut off a valve upstream from the leak. (If you think it will take you a while to locate and turn off this valve, direct the leak into a suitably sized container to minimize damage until you can turn the water off.)
- Remove excess water.
- Move furniture, equipment, or other obstacles clear of the work area.
- Cover furniture, floors, and equipment to protect them from any damage that could occur as a result of your work on the valve.
- Place ladders carefully, bracing them if necessary.
- Turn the water back on to test your repair.
- Do your work neatly and clean up when you finish.

10.2.0 Identifying and Repairing Defects

After you've taken the necessary safety precautions, inspect the valve. The defect is related to the type of valve. Valves come in many styles, but for repair work, they fit into the following five categories:

- Globe valves
- Gate valves
- Flushometers
- Flush valves
- Float-controlled valves/ball cocks

10.2.1 Repairing Globe Valves

Internally, globe valves, angle valves, and compression faucets contain the same basic parts (see *Figure 30*). Therefore, the problems usually encountered and their solutions are similar (see *Table 1*).

To prevent damage to the finished surfaces of the valves, use the correct size wrenches. Don't use pliers or adjustable wrenches. Make sure the wrench jaws are clean and smooth to prevent scratching.

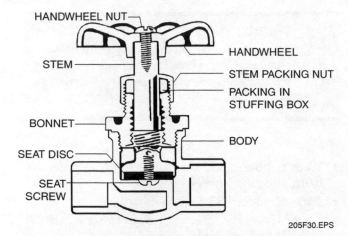

Figure 30 ◆ Basic parts of globe and angle valves and compression faucets.

Table 1 Troubleshooting Globe and Angle Valves and Compression Faucets

Problems	Possible Causes
1. Drip or stream of water flows when valve is closed	Worn or damaged seat Worn or damaged seat disc
2. Leak around stem or from under knob	Loose packing nut Defective packing Worn stem
3. Rattle when valve is open and water is flowing	Loose seat disc Worn threads on stem
4. Difficult or impossible to turn handwheel or knob	Packing nut too tight Damaged threads on stem

10.2.2 Repairing Gate Valves

Some of the problems common to gate valves are similar to those affecting globe valves (see *Figure 31*). Leaks around the stem are the result of either packing wear or wear on the stem.

Gate valves must operate either fully opened or fully closed; don't use them to throttle the volume of liquid flowing through the pipe. When gate valves open only part way, the disc, or gate, vibrates. This vibration erodes the disc edge, a defect commonly called **wire drawing**. You can't repair this valve to solve the problem; instead, replace it with a ball valve or globe valve.

If a gate valve fails to stop the flow of water, the cause is one of three possible problems:

- The disc, or gate, is worn.
- The valve seat is worn.
- Some foreign material is preventing the disc from seating properly.

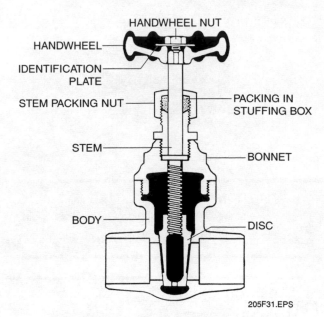

Figure 31 ◆ Basic parts of the gate valve.

To determine which of these problems exists, remove the bonnet and inspect the gate and seat. You can try to solve the problem by scraping out mineral deposits and cleaning the mating surfaces, but you may have to replace the disc. If the seat area is worn, replace the valve. Note that some companies remanufacture larger gate valves. You may exchange gate valves 2 inches in diameter and larger for remanufactured gate valves.

DID YOU KNOW?

The early 1920s saw the introduction of the first quarter-turn plug valve. This valve operated with a simple 90-degree turn of the handle. Plug valves found wide application in the chemical and gas industries.

During World War II, a British army officer invented the diaphragm valve. This tight, corrosion-resistant valve featured a soft rubber disc bolted between the body and the bonnet. World War II presented a special challenge to the valve industry and the U.S. Navy. Concussions from bombs dropped close to ships fractured many of the standard valves on board. The Navy replaced thousands of standard valves with impact-resistant valves designed to handle the shock waves.

10.2.3 Repairing Flushometers

These are four common problems associated with flushometers:

- Leakage around the handle
- Failure of the vacuum breaker
- Control stop malfunction
- Leakage in the diaphragm that separates the upper and lower chambers

Several manufacturers produce flushometers and sell kits containing repair components. All flushometers contain the same basic components; however, specific parts may vary. Follow the manufacturer's specifications when installing replacement parts. The troubleshooting guide in *Table 2* applies to flushometers.

10.2.4 Repairing Float-Controlled Valves/Ball Cocks

To repair a malfunctioning valve, use one of the repair kits available from the manufacturer. These kits contain replacement parts. If you find damage to the stem, valve body, or float mechanism, replace the entire float valve.

10.2.5 Repairing Flush Valves

If you find severe corrosion of the valve or the lever that activates it, replace both parts. The more common problems, however, involve the component parts. Inspect the tank ball (or flapper tank ball), the chains (or wires), and the guide for deterioration. Depending on the condition of these components, make repairs or replace the entire assembly. Use a reseating tool to restore a corroded valve seat.

Table 2 Troubleshooting Guide for Flushometers

Problem	Cause	Solution
1. Nonfunctioning valve	Control stop or main valve closed	Open control stop or main valve
2. Not enough water	Control stop not open enough Urinal valve parts installed in closet parts Inadequate volume or pressure	Adjust control stop to siphon fixture Replace with proper valve Increase pressure at supply
3. Valve closes off	Ruptured or damaged diaphragm	Replace parts immediately
4. Short flushing	Diaphragm assembly and guide not hand-tight	Tighten
5. Long flushing	Relief valve not seating	Disassemble parts and clean
6. Water splashes	Too much water is coming out of faucet	Throttle down control stop
7. Noisy flush	Control stop needs adjustment Valve may not contain quiet feature The water closet may be the problem	Adjust control stop Install parts from kit Place cardboard under toilet seat to separate bowl noise from valve noise—if noisy, replace water closet
8. Leaking at handle	Worn packing Handle gasket may be missing Dried-out seal	Replace assembly Replace Replace

Review Questions

Sections 7.0.0–10.2.5

1. _____ stems are suitable for areas where space is limited.
   a. Rising
   b. Nonrising
   c. OS&Y
   d. Sliding

2. The three basic types of valve bonnets are _____.
   a. union bonnet, bolted bonnet, flanged bonnet
   b. bolted bonnet, gate bonnet, globe bonnet
   c. screwed bonnet, union bonnet, bolted bonnet
   d. pressure bonnet, supply stop bonnet, union bonnet

3. The most common end connection in plumbing systems is the _____ end.
   a. solder
   b. flanged
   c. internal thread
   d. external thread

4. Repair of a gate valve that vibrates or shows signs of wire drawing is best done by _____.
   a. replacing the disc in the current valve
   b. replacing the stem in the current valve
   c. installing a new ball valve or globe valve
   d. installing a new gate valve with a larger disc

5. A common problem associated with flushometers is _____.
   a. a deteriorated chain
   b. a corroded valve or lever
   c. a cracked tank ball
   d. leakage around the handle

Summary

Valves regulate flow. They may provide on/off service, act as a throttling device, or prevent flow reversal through a line. The six types of valves are gate, globe, angle, ball, butterfly, and check. Materials used to make valves include iron, brass, carbon steel, bronze, aluminum, alloy steels, and polyvinyl chloride—practically every machinable material.

The four basic types of stems on valves are rising, nonrising, outside screw and yoke, and sliding. Packing material provides a tight seal around the stem. Plumbers join valves, like pipes, in several ways. The four basic types of end connections are internal threads, external threads, solder ends, and flanged ends.

Valves come in many styles, but for repair work, they fit into five categories: globe valves, gate valves, flushometers, flush valves, and float-controlled valves/ball cocks. The repair methods for globe and gate valves are similar. Manufacturers provide repair kits containing replacement parts for flushometers, flush valves, and float-controlled valves/ball cocks.

When repairing valves, think safety first. Take steps to protect yourself and others on the job site and to minimize property damage.

Trade Terms Introduced in This Module

Angle valve: A valve that is similar to the globe valve but that can serve as both a valve and a 90-degree elbow.

ASME Power Boiler Vessel Code: A standard produced by the American Society of Mechanical Engineers that sets blowdown limitations for steam safety valves.

Ball valve: A valve that regulates fluid flow with a movable ball that fits in a sphere-shaped seat.

Ball-check valve: A valve that allows one-way flow in water supply or drainage lines.

Blowdown: The difference between the opening and closing pressures in a steam safety valve.

Bolted bonnet: A valve cover that guides and encloses the stem. It is used for larger size valves or for high-pressure applications.

Bonnet: A valve cover that guides and encloses the stem.

Butterfly valve: A valve that controls the flow of fluids. Two discs are hinged on a common shaft to permit flow in only one direction.

Check valve: An automatic valve that permits flow of liquid in only one direction.

External thread: A valve end connection that has the threads on the outside.

Flanged end: A valve end connection that is used in areas where frequent dismantling of the pipe is necessary.

Float-controlled valve: A valve installed in water closet flush tanks that maintains a constant water level in the tank.

Flush valve: A valve installed in a water closet flush tank that controls the flow of water from the tank into the bowl.

Flushometer: A valve that allows a preset amount of water to enter a fixture such as a water closet or urinal to flush it.

Full flow: The relative flow capacity of a particular valve.

Gate valve: A flow-control device with a gate, or disc, that can be raised or lowered. It is not intended for close fluid flow control or for very tight shutoff.

Globe valve: A valve that controls the flow of liquid with a movable spindle, which lowers to a fixed seat to restrict flow through the valve opening.

Huddling chamber: A chamber within a safety valve that harnesses the expansion forces of air or steam to quickly open (pop) or close the valve.

Internal thread: A valve end connection that has the threads inside.

Lever-and-weight valve: A type of swing-check valve used when pulsation occurs in the line.

Lift-check valve: A type of valve that prevents flow reversal in a piping system. It is used for gas, water, steam, or air and for lines where frequent fluctuations in flow occur.

Nonrising stem: A type of stem in which neither the handwheel nor the stem rises when the valve is opened. It is suitable for areas where space is limited.

Outside screw and yoke stem: A type of valve stem suitable for use with corrosive fluids because it does not come in contact with the fluid line. Often abbreviated OS&Y.

Packing: A material that seals the valve stem and prevents fluids from leaking up through it.

Pressure regulator valve: A valve that reduces water pressure in a building.

Rising stem: A type of valve in which both the handwheel and the stem rise.

Screwed bonnet: A two-piece unit that is used in low-pressure applications and where frequent disassembly of the valve is not required.

Sliding stem: A type of valve stem suitable for applications where quick opening and closing is required.

Slurries: Liquids that contain large amounts of suspended solids.

Solder end: A type of end connection for valves. The smooth ends are soldered to copper and brass pipes that have no internal or external threads.

Solid-wedge valve: A type of gate valve that is used for steam, hot water, and other services where shock is a factor.

Split-wedge valve: A type of gate valve that is used on lines that require a more positive closure—for example, low-pressure and cold water lines. It is also used on lines containing volatile fluids.

Straight-through flow: An unrestricted flow of fluids through a pipe.

Supply stop valve: A valve used to disconnect the hot or cold water supply to individual fixtures such as water closets, lavatories, and sinks. Also called the *supply stop.*

Swing-check valve: A type of check valve that features a low-flow resistance, making it suited for lines containing liquids or gases with low to moderate pressure.

Temperature/pressure relief valve: A valve that limits fluid pressure by discharging some of the pressurized liquid or gas. Abbreviated T/P valve.

Throttled flow: A valve designed to partially control or throttle the flow of liquid.

Trim: The parts of a valve that receive the most wear and tear: the stem, disc, seat ring, disc holder, wedge, and bushings.

Union bonnet: A three-piece unit best suited for those applications where frequent disassembly of the valve is required.

Water hammer: In water lines, a loud thumping noise that results from a sudden stoppage of the flow. In steam lines, water or condensation is carried through the steam main at high speed; when the direction of flow changes, water particles hit pipe walls with a banging sound.

Wire drawing: A condition caused when gate valves open only part way causing the disc, or gate, to vibrate, which in turn erodes the disc edge.

Answers to Review Questions

Sections 2.0.0–3.6.0
1. a
2. d
3. b
4. a
5. c

Sections 4.0.0–6.0.0
1. b
2. d
3. b
4. c
5. a

Sections 7.0.0–10.2.5
1. b
2. c
3. a
4. c
5. d

NCCER CRAFT TRAINING USER UPDATES

The NCCER makes every effort to keep these textbooks up-to-date and free of technical errors. We appreciate your help in this process. If you have an idea for improving this textbook, or if you find an error, a typographical mistake, or an inaccuracy in the NCCER's Craft Training textbooks, please write us, using this form or a photocopy. Be sure to include the exact module number, page number, a detailed description, and the correction, if applicable. Your input will be brought to the attention of the Technical Review Committee. Thank you for your assistance.

Instructors – If you found that additional materials were necessary in order to teach this module effectively, please let us know so that we may include them in the Equipment and Materials list in the Instructor's Guide.

Write: Curriculum Revision and Development Department
National Center for Construction Education and Research
P.O. Box 141104, Gainesville, FL 32614-1104

Fax: 352-334-0932

E-mail: curriculum@nccer.org

Craft _____ Module Name _____

Copyright Date _____ Module Number _____ Page Number(s) _____

Description _____

_____

_____

_____

(Optional) Correction _____

_____

_____

(Optional) Your Name and Address _____

_____

_____

Installing and Testing
Water Supply Piping

COURSE MAP

This course map shows all of the modules in the second level of the Plumbing curriculum. The suggested training order begins at the bottom and proceeds up. Skill levels increase as you advance on the course map. The local Training Program Sponsor may adjust the training order.

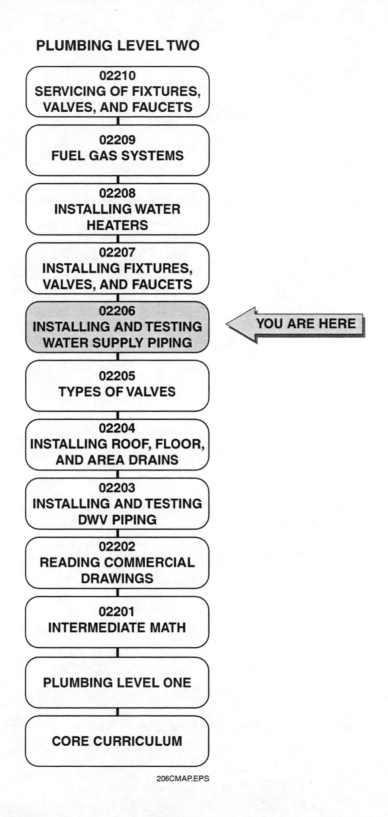

PLUMBING LEVEL TWO

02210
SERVICING OF FIXTURES,
VALVES, AND FAUCETS

02209
FUEL GAS SYSTEMS

02208
INSTALLING WATER
HEATERS

02207
INSTALLING FIXTURES,
VALVES, AND FAUCETS

02206
INSTALLING AND TESTING
WATER SUPPLY PIPING

← YOU ARE HERE

02205
TYPES OF VALVES

02204
INSTALLING ROOF, FLOOR,
AND AREA DRAINS

02203
INSTALLING AND TESTING
DWV PIPING

02202
READING COMMERCIAL
DRAWINGS

02201
INTERMEDIATE MATH

PLUMBING LEVEL ONE

CORE CURRICULUM

206CMAP.EPS

Figures

Table

Installing and Testing Water Supply Piping

Objectives

When you have completed this module, you will be able to do the following:

1. Develop a material takeoff from a given set of plans.
2. Use plans and fixture rough-in sheets to determine the location of fixtures and the route of the water supply piping.
3. Demonstrate the ability to locate and size a water meter.
4. Demonstrate the ability to locate a water heater, water softener, and hose bibbs.
5. Demonstrate the ability to install a water distribution system using appropriate hangers.
6. Modify structural members, using the appropriate tools, without weakening the structure.
7. Demonstrate the ability to safely size and install a water service line and provide for water hammer protection.
8. Demonstrate the ability to test a water supply system.

Prerequisites

Before you begin this module, it is recommended that you successfully complete the following modules: Core Curriculum; Plumbing Level One; Plumbing Level Two, Modules 02201 through 02205.

Required Trainee Materials

1. Paper and sharpened pencil
2. Appropriate Personal Protective Equipment

1.0.0 ◆ INTRODUCTION

Water supply piping is an important concern in residential as well as commercial buildings. Customers expect—and should have—adequate water at fixtures, quiet pipes, pipes that don't freeze, and clean water that tastes good. Careful design and proper installation will ensure the best possible service from a building's water supply system.

In this module, you'll learn about the general design and installation concepts used for water supply systems. The potable (drinkable) water supply system includes the water service pipe, the water meter, the water-distributing pipe, connecting pipes, pipe hangers and supports, fittings, control valves, and all other necessary components located in or next to the building.

2.0.0 ◆ PLANS

For economy and efficiency, plumbers work according to plans or blueprints. For most industrial, commercial, and multi-residential structures, plumbing plans are included with the architectural plans for the structure. These plans are designed with a certain amount of flexibility to allow the plumber to make on-the-job decisions when confronted by unusual circumstances. In many cases, particularly in light residential applications, plumbing plans are not included with the prints. In these cases, the plumber uses the floor plan to locate the runs of pipe and the various plumbing fixtures (see *Figure 1*).

2.1.0 Material Takeoffs

A material takeoff is a list of the type and quantity of material required for a given job. An estimator who is knowledgeable about plumbing practices and materials may work from a set of drawings to do the takeoff. There is no standard form used for material takeoffs. A typical form is shown in *Figure 2*.

3.0.0 ◆ MAIN TO METER WATER SERVICE

In some areas, the water service piping begins at the utility water main and either connects into the main itself or starts on the outlet side of a corporation stop provided by the water utility. In other areas, the water utility brings the water supply to the property line and terminates it with a **curb stop**. A **curb box** provides access to the supply (see *Figure 3*).

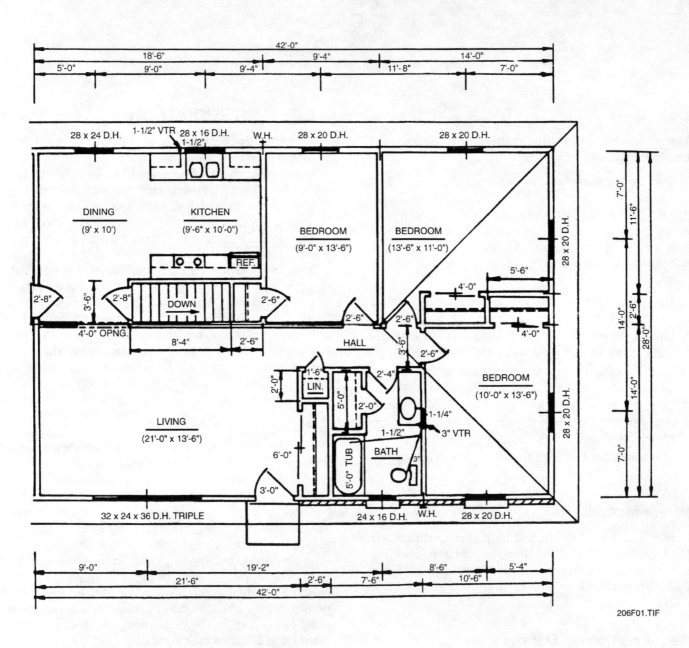

206F01.TIF

Figure 1 ◆ House floor plan.

MATERIALS ESTIMATE

PROJECT _Joe Doe — Residence_   ESTIMATE NO. _107_

LOCATION _Anywhere, U.S.A._   SHEET NO. _1_

ARCHITECT ENGINEER _John Smith_   DATE _10-2-81_

SUMMARY BY _E.K.B._   PRICES BY _P.J._   CHECKED BY _M.B._

QUANTITY	DESCRIPTION	UNIT PRICE	TOTAL ESTIMATED MATERIAL COST	UNIT PRICE	TOTAL ESTIMATED LABOR COST	TOTAL
2	19" Round Lavs.	21 ⁹⁵	43.90	18/hr.	90.00	133 90
2	Closets	51 ⁰⁰	102.00	"	72 00	174 00
1	Tub-Shower Value	22 ⁸⁶	22.86	"	12 00	34 86
1	Kitchen Sink	56 ¹⁰	56.10	"	54.00	110 10
20'	4" ABS Pipe	1 ²³ FT.	24 60			24 60
80'	1½" ABS Pipe	.31 FT.	24 80			24 80
4	4" TY	4 ²³	16.92			16 92
8	2" 90° Ells	.30 EA.	2.40			2 40
2	3"×1½" Tees	.96 EA.	1.92			1 92
2	3" Flashings	8 ⁶⁵	17 30			17 30

206F02.TIF

Figure 2 ◆ Material takeoff.

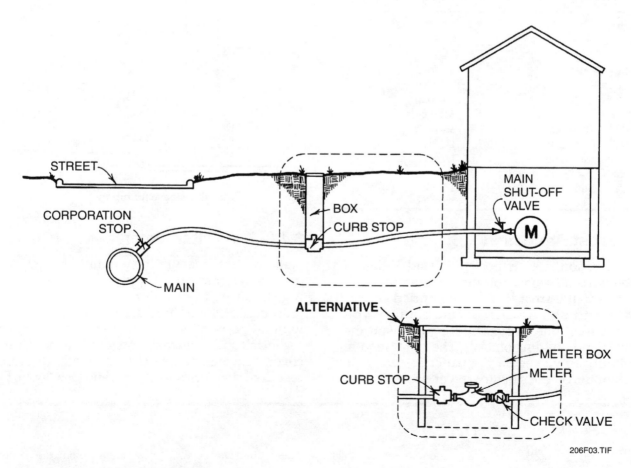

206F03.TIF

Figure 3 ◆ Curb stop and curb box.

3.1.0 Piping Material and Sizes

Consult the local plumbing code to learn which piping materials are approved. Copper tubing and certain plastics are the materials most commonly used.

To determine the size of the water supply piping, you must know the demand for the proposed building. Factors to consider include the following:

- The type of flush devices on the fixtures
- Water pressure in pounds per square inch (psi) at the source
- The length of pipe in the building
- The number and kinds of fixtures installed
- The number of fixtures expected to be used at any time

Table 1 shows recommended pipe sizes for typical residential fixtures. As you can see, most pipe sizes average between ½ and ¾ inch.

Table 1 Recommended Pipe Sizes for Residential Fixtures

Fixture	Pipe Size
Bathtub	1/2 inch
Dishwasher	1/2 inch
Hose Bibb	3/4 or 1/2 inch
Water Heater	1/2 inch
Kitchen Sink	1/2 inch
Lavatory	1/2 inch
Shower	1/2 inch
Urinal – Flush Valve	3/4 inch
Urinal – Flush Tank	1/2 inch
Washing Machine	1/2 inch
Water Closet – Flush Valve	1 inch
Water Closet – Flush Tank	3/8 inch

3.2.0 Frost Protection

Frost protection is an important consideration in plumbing installations. You must maintain a safe depth below the **frost line** for the buried service line. Depending on where in the country you work, the frost line depth will vary. Consult the local code and building officials. The codes require accessibility to valves on the water service pipe and cleanouts on the sanitary sewer drainpipe.

In colder climates, locate the service entrance in areas other than under sidewalks, driveways, and patios. Snow acts like insulation and keeps frost from penetrating the soil too deeply. Therefore, the frost depth on snow-covered ground is significantly less than it is in areas that are routinely cleared of snow, like driveways. Note also that in wet soils frost penetrates deeper than it does in dry soils. If possible, locate water pipes under lawns or other areas where snow will remain.

3.3.0 Pipe Protection

Place sleeves around pipes that pass through a building's structural concrete—for example, floors, footings, stem walls, and grouted masonry (see *Figure 4*). Sleeves protect the pipe from damage by contact with concrete. They are made from foam or plastic pipe that is one size larger than the pipe they protect. Secure sleeves in place before pouring the concrete, and slope them to prevent pinching on the pipe.

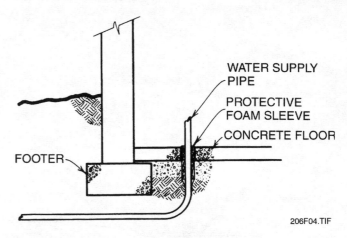

WATER SUPPLY PIPE
PROTECTIVE FOAM SLEEVE
CONCRETE FLOOR
FOOTER

206F04.TIF

Figure 4 ◆ Protective sleeve for piping.

3.4.0 The Water Meter

Water meters measure the amount of water passing through the water supply piping into a building. Usually, they belong to a city or a local municipality, and you may not remove them without permission. Water department employees install and service the meters. The housing, or cover, of water meters may be made of bronze, cast iron, stainless steel, or plastic (see *Figure 5*).

ON THE ·LEVEL·

Friction

The greater the number of fittings in the system (valves, tees, elbows, and so on), the greater the amount of friction, which in turn causes a pressure drop in the system.

The amount of water used varies for each customer. Meters are made with measurement capacities of 20, 30, or 50 gallons per minute. While the
interior design varies from meter to meter, most
meters have a magnetic drive.

As the plumber, it is your responsibility to
plumb the meter. Most utility companies will set
meters up to 1½ inches in diameter. Plumbers usually plumb and set meters that require a larger
pipe diameter. Leave the proper spacing between
pipes to accommodate a water meter.

When a water meter is located inside a building, you can install an external meter register to
make meter reading easier (see *Figure 6*).

Install a main shutoff valve to control the water
flow during and after construction. Local codes
vary as to the type of valve required. You can
locate the water meter directly after the main
shutoff valve or before the main shutoff valve.

4.0.0 ◆ WATER HEATER, SOFTENER, AND HOSE BIBBS

The working drawings will show the location of
the water heater, the water softener, and **hose
bibbs**. However, you have a certain amount of
flexibility in locating these items. Each building
presents its own challenges, and you may need to
think creatively to resolve them or to make the
system more efficient.

206F05.EPS

Figure 5 ◆ Water meters.

206F06.EPS

Figure 6 ◆ External meter register.

ON THE · LEVEL ·

Before and After

The meter has an inlet side and an outlet side. The inlet side receives water from the street
main, and the outlet side sends water to the building. So a location before the meter is on the
inlet side, and a location after the meter is on the outlet side. Some plumbers use the terms
upstream instead of before and downstream instead of after.

4.1.0 Locating the Water Heater

To place the water heater in the most efficient location, minimize the length of the piping runs. To do this, ensure that the greatest number of hot water outlets are as close as possible to the hot water heater.

A customer may want to provide for solar hot water heating at some point in the future. If so, it's a good idea to include fittings during installation that will make the transition easier. Although some solar storage tanks are one and the same with the water heater, other systems require two tanks for more efficient operation. In this case, provide an enclosure large enough to accommodate the desired system during the construction phase to eliminate unnecessary rebuilding later on. Depending on the job specifications, the plumber may construct the enclosure or may subcontract this job to the general contractor.

4.2.0 Locating the Water Softener

Before installing the water supply piping, determine where in the building treated water is required. The plans may specify this, or you may need to consult with the customer. For example, residential hot water usually is softened. The water supplied to hose bibbs, on the other hand, may bypass the softener. Depending on the quality of the drinking water, a customer may prefer to have treated water at the kitchen sink's cold water tap. The rest of the cold water outlets in the home may or may not require softening.

If you run a drain line from the softener to the floor drain, make certain that you maintain an appropriate air gap to prevent backflow.

4.3.0 Locating the Hose Bibbs

At least one exterior water outlet, or **hose bibb**, is required in residential structures, and the customer may want more. These runs are usually long and normally bypass the water softener. Hose bibbs are usually ¾ inch, although you may use ½-inch piping to access the outlet.

Review Questions

Sections 2.0.0–4.3.0

1. The type and quantity of plumbing supplies needed for a job are listed by an estimator on the _____.
   a. architect's blueprint
   b. elevation drawing
   c. material takeoff
   d. building plan

2. The location of a water supply line and the depth below the frost line for a single-family residence usually are determined by the _____.
   a. water pressure
   b. local building codes
   c. distance from the curb stop
   d. number of fixtures in the building

3. In cold climates, it is best to locate water service under _____.
   a. sidewalks
   b. driveways
   c. snow-covered grassy areas in drier soils
   d. snow-covered grassy areas in wetter soils

4. The _____ of a water meter sends water to the building after it passes through the meter.
   a. outlet side
   b. inlet side
   c. outdoor side
   d. indoor side

5. To place a water heater in the most efficient location, you should _____.
   a. maximize the length of the piping runs
   b. choose a southern corner of the building
   c. limit the number of hot water outlets near the heater
   d. minimize the length of the piping runs

ON THE · LEVEL ·

Fixture Rough-In

Refer to *Figure 7.* It shows an example of fixture rough-in measurements. You will find these measurements in manufacturers' catalogs, which are available from plumbing distributors. Note that critical dimensions may vary between styles of the same brand (refer to A and B in the figure). Rough-in measurements tell you exactly where the piping should exit the walls or floors. You will use these dimensions to run the water supply piping and to attach the fixtures to the floor or wall.

5.0.0 ◆ LOCATING THE FIXTURES

Note that before you install the water supply piping, the drain, waste, and vent (DWV) piping is already in place. Because DWV piping is large, and because plumbers have less flexibility with it, they install the DWV first. So you must locate the water supply piping in reference to the DWV piping. Although some flexibility exists with the water supply pipe location, you should make every effort to locate the water supply as accurately as possible.

Install the water supply piping in reference to the drains. See *Figure 7*, which gives reference dimensions for proper pipe location. Fixture manufacturers have these specifications. Note that specifications apply to the *style* of fixture selected. For example, if you refer to illustrations (A) and (B) in *Figure 7*, you'll see that they are identical except for the measurement of the centerline to the drain. Notice also that the water supply is referenced from the drain. If you don't locate the drain properly, you may have connection problems (depending on the fixture). When installing water supply piping, it is a good idea to recheck the location of the drains to ensure correct positioning.

Notice that for fixtures such as water closets, references are made to the *finished* wall surface. The thickness of interior wall coverings—drywall, plaster, wood, or ceramic tile—varies. Locate the drain with the final wall thickness in mind.

ON THE

·LEVEL·

Preventing Corrosion and Backflow

When different metals—like copper and iron—come into contact with one another in the presence of water, they can corrode. A **dielectric fitting** is a special type of adapter used to connect pipes of different metals. The adapter prevents corrosion from taking place. To prevent damage to the water supply system, use dielectric fittings on water heaters, water softeners, and solar collectors.

Backflow occurs when liquids flow into the water supply's distributing pipes or into any fixture or appliance from a source opposite to the intended flow. Install backflow preventers in areas where hoses can suck contaminated water into the potable water system. These areas include hose bibbs, dishwashers, and slop sinks.

Dielectric fittings and backflow prevention devices are not always included in codes or specifications. Knowing about them enables you to offer efficient and safe service to customers.

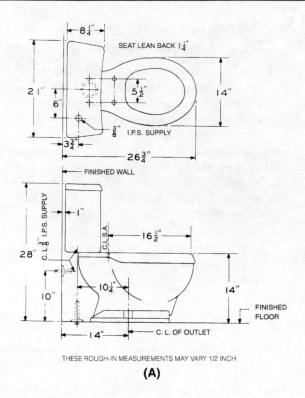

THESE ROUGH-IN MEASUREMENTS MAY VARY 1/2 INCH

(A)

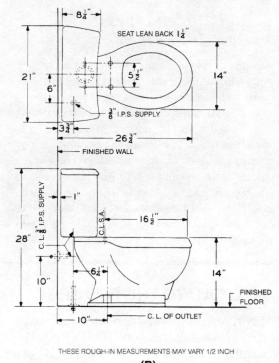

THESE ROUGH-IN MEASUREMENTS MAY VARY 1/2 INCH

(B)

VITREOUS CHINA COUNTERTOP LAVATORY   CENTERSET FITTING

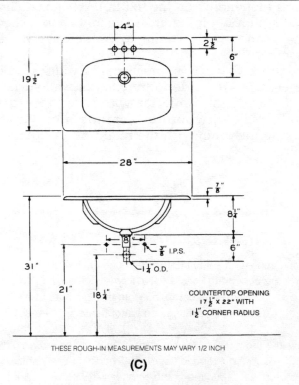

THESE ROUGH-IN MEASUREMENTS MAY VARY 1/2 INCH

(C)

OVER-RIM BATH FILLER
BATH AND SHOWER FITTING WITH DIVERTER VALVE
BATH AND SHOWER FITTING WITH DIVERTER SPOUT
512-1150 POP-UP BATH WASTE

ENAMELED FORMED STEEL BATH

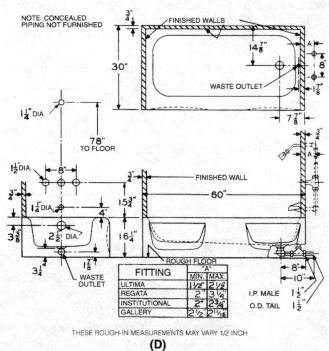

FITTING	"A"	
	MIN.	MAX.
ULTIMA	1½"	2½"
REGATA	2"	3¼"
INSTITUTIONAL	2"	2¾"
GALLERY	2½"	2¹¹/₁₆"

THESE ROUGH-IN MEASUREMENTS MAY VARY 1/2 INCH

(D)

206F07.TIF

Figure 7 ◆ Locating fixtures.

5.1.0 Assembling and Installing the Stubouts

The stubout is that part of the pipe that extends beyond the finished wall. Once the wall is finished, you'll install the angle stop and the flexible fixture riser. The riser extends from the angle stop and connects to the fixture faucet or common tank.

After you have assembled, capped, and soldered the stubouts, place the assembly through the access hole in the floor to the feeder lines below (see *Figure 8*).

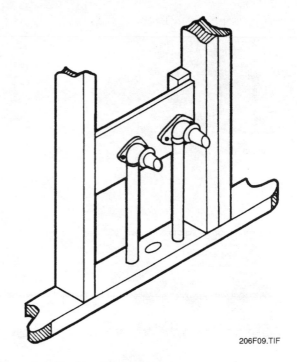

206F09.TIF

Figure 9 ◆ Backing board.

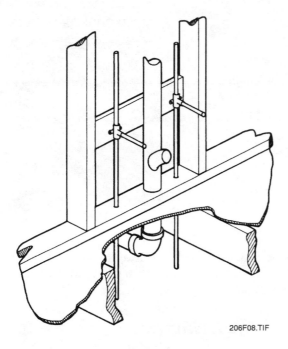

206F08.TIF

Figure 8 ◆ Installation of fixture stubouts.

With the risers visible below the floor, you can easily locate the fixtures from below. To position the riser and stubout assemblies firmly in their permanent location, use a backing board to mount the drop ells (see *Figure 9*). A drop ear ell and a high ear ell are shown in *Figure 10*. The ears on these fittings serve to anchor the pipe in place on the inside of the finished wall. Check the job specifications to determine which fitting to use.

6.0.0 ◆ INSTALLING PIPE HANGERS AND SUPPORTS

You must install and support all water distribution pipe so that both the pipe and its joints remain leakproof. Improper support can cause the joints to sag. This causes added stress on the pipe and increases the probability of leaks, breaks, or even cracks between the joints.

206F10.EPS

Figure 10 ◆ Drop ear ell and high ear ell.

SUSPENSION HANGER

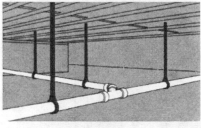

STRAP LOCK HANGER

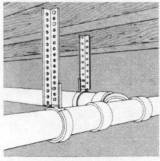

MULTI-PURPOSE HANGER

PLASTIC PLUMBERS TAPE

SNAP STRAP

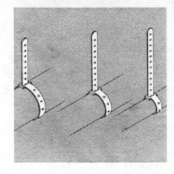

MULTI-PURPOSE HANGER

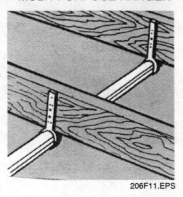

206F11.EPS

Figure 11 ◆ Plastic hangers and supports.

6.1.0 Types of Pipe Hangers and Supports

Supports and hangers are designed to hold and support pipe in either a horizontal or a vertical position (see *Figure 11*). They are made from various materials, including carbon steel, malleable iron, steel, cast iron, and plastic. They come in different finishes, such as copper plate, black, and electrogalvanized.

Which hanger style and finish to use depends upon the type of pipe and its application. Because various hangers and supports are used in combination with each other, it is difficult to set up naturally exclusive categories. However, the basic components of hangers and supports can be placed into three major categories. These are pipe attachments, connectors, and structural attachments.

6.1.1 Pipe Attachments

Pipe attachments include the part of the hanger that touches or connects directly to the pipe. They may be designed for either heavy duty or light duty, for covered pipe or plain pipe.

Several styles of hangers used for supporting piping from the ceiling horizontally are illustrated in *Figure 12*. These include various types of rings, clamps, and clevises. The styles illustrated are recommended for suspending both hot and cold noninsulated stationary piping.

CLEVIS HANGER

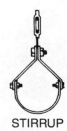
STIRRUP

ADJUSTABLE SWIVEL RING HANGER

206F12.EPS

Figure 12 ◆ Hangers used to support pipe from the ceiling horizontally.

You can support piping on wood frame construction with several styles of pipe attachments. These include pipe hooks, U-hooks, J-hooks, tube straps, tin straps, perforated band irons, half-clamps, suspension-clamps, and pipe clamps (see *Figures 13* and *14*).

Several styles of pipe attachments are available for supporting piping from the walls or the sides of beams and columns. *Figures 15* and *16* illustrate several styles of pipe attachments used for this purpose. They include one-hole clamps, steel brackets, offset clamps, various styles of clips and straps, and combinations using extension split-clamp hangers and wall plates.

Support vertical pipes at each floor level using riser clamps (see *Figure 17*). These clamps are available in different finishes. Mount the riser clamp so that the bracket supports the pipe weight directly on the floor.

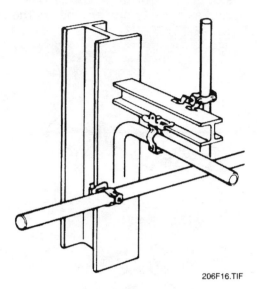

Figure 16 ◆ Clip-type attachments for beams and columns.

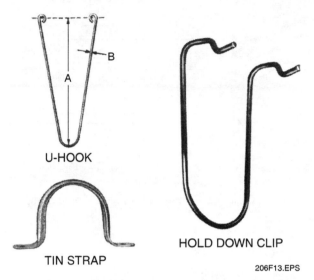

Figure 13 ◆ Pipe supports for wood-frame construction.

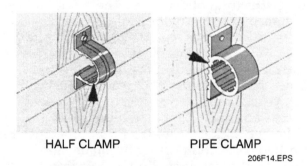

Figure 14 ◆ ABS plastic pipe clamps for wood.

Figure 17 ◆ Riser clamps.

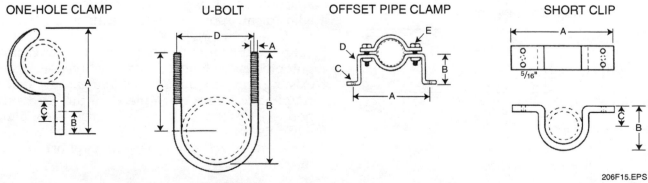

Figure 15 ◆ Attachments for use on walls or beams and columns.

Other pipe attachments include universal pipe clamps (see *Figure 18*) and standard 1⅝-inch or 1½-inch channels (see *Figure 19*). The clamps are available in standard finishes of mild and electro-galvanized steel. Aluminum, copper-plated, and stainless steel finishes are also available from some manufacturers, but usually these finishes have to be special-ordered. The notched steel clamps are inserted by twisting them into position along the slotted side of the channel. The pipes can be aligned as close to one another as the couplings allow.

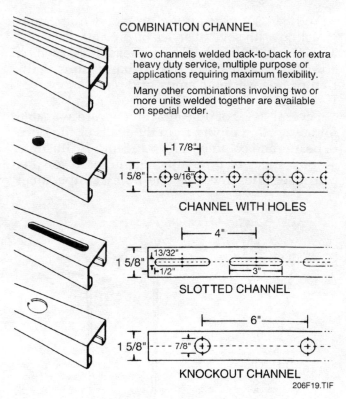

Figure 19 ◆ Channels for use with universal pipe clamps.

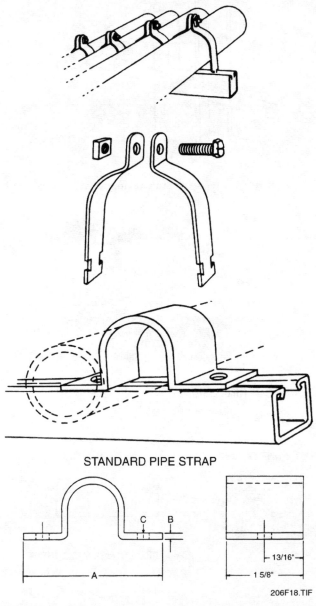

Figure 18 ◆ Universal pipe clamps.

6.1.2 Connectors

The connector portion of the hanger is the intermediate attachment that links the pipe attachment to the structural attachment. These intermediate attachments can be divided into two groups: rods and bolts, and other rod attachments.

The rod attachments include eye sockets, extension pieces, rod couplings, reducing rod couplings, hanger adjusters, turnbuckles, clevises, and eye rods. All are available in several sizes that will meet most installation requirements.

The eye socket provides for a nonadjustable threaded connection. Use it in conjunction with a hanger rod when installing a split-ring hanger. Extension pieces are used for attaching hanger rods to beam clamps and other types of building attachments. They provide for a small amount of adjustment, approximately 1 inch, on the hanger rod (see *Figure 20*).

Rod couplings and reducing rod couplings support pipelines where it is possible to connect to an existing stud (see *Figure 21*). You can use either coupling to connect two pieces of threaded support rod. However, use the reducing rod coupling when the support rods are of different sizes and the standard rod coupling with support rods of the same size.

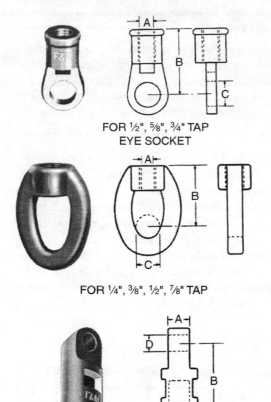

FOR ½", ⅝", ¾" TAP
EYE SOCKET

FOR ¼", ⅜", ½", ⅞" TAP

EXTENSION PIECE

206F20.EPS

Figure 20 ◆ Eye sockets and extension piece.

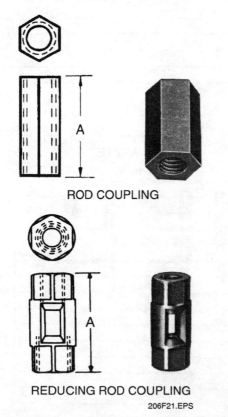

ROD COUPLING

REDUCING ROD COUPLING

206F21.EPS

Figure 21 ◆ Rod coupling and reducing rod coupling.

Hanger adjusters and turnbuckles provide an adjustable threaded connection for the support rod (see *Figure 22*).

You can use hanger adjusters with split-ring hangers or beam clamps when an adjustable hanger rod connection is desirable. Hanger adjusters have a permanently set swivel in the body that allows smooth adjustment during installation. Turnbuckles are used for connecting two hanger support rods. They provide up to 6 inches of adjustment.

Use the weldless eye nut and the forged steel clevis on high-temperature piping installations (see *Figure 23*). Install the eye nut where a flexible connection is required. Use a clevis to connect the support rod to the welded lug or structural steel on heavy-duty piping installations.

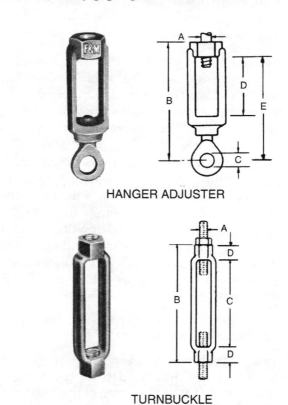

HANGER ADJUSTER

TURNBUCKLE

206F22.EPS

Figure 22 ◆ Hanger adjuster and turnbuckle.

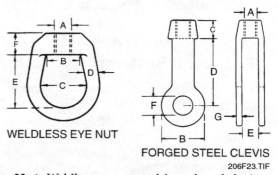

WELDLESS EYE NUT

FORGED STEEL CLEVIS

206F23.TIF

Figure 23 ◆ Weldless eye nut and forged steel clevis.

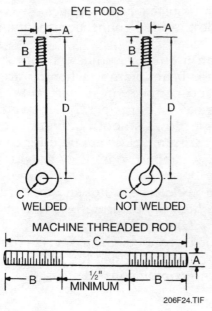

EYE RODS

WELDED   NOT WELDED

MACHINE THREADED ROD

½"
MINIMUM

206F24.TIF

Figure 24 ◆ Hanger rods.

The hanger rod portion of the connector includes the eye rods and the machine threaded rods (see *Figure 24*). They are used to connect the rod attachments to the pipe attachments, forming a hanger assembly. The eye rods are available without a welded eye for lighter installations and with a welded eye for installations where more strength is required. Machine threaded rods are available with continuous threads that run the complete length. You can cut them to the required hanger length on the job, eliminating the need for field threading. Hanger rods are available with two threaded ends—with right-hand threads on both ends or with one right-hand thread and one left-hand thread.

6.1.3 Structural Attachments

Structural attachments are the anchors and anchoring devices that hold the pipe hanger assembly securely to the structure of the building. These include powder-actuated anchors, concrete inserts, beam clamps, C-clamps, beam attachments, various brackets and ceiling flanges, plates, plate washers, and lug plates.

6.2.0 Powder-Actuated Fastening Systems

The use of powder-actuated anchor systems has increased rapidly in recent years. They are widely approved for anchoring static loads (stationary and vibration-free) to steel and concrete beams, walls, and so forth. Consult agencies such as Underwriter's Laboratories (UL) and local building codes for specific types of loads and load lim-

WARNING!

You must wear appropriate personal protective equipment when using powder-actuated fastening tools. Never point these tools at anyone. Serious injury or death could result. Be sure to read and follow the manufacturer's safety instructions.

its allowable in your area. In addition, requirements of the Occupational Safety and Health Administration (OSHA) specify that plumbers using these systems must be certified in their operation. Certification can be arranged through factory sales representatives.

Powder-actuated tools drive steel pins or threaded steel studs directly into cured concrete or structural steel surfaces to hold pipe hangers, brackets, and so on (see *Figures 25* and *26*).

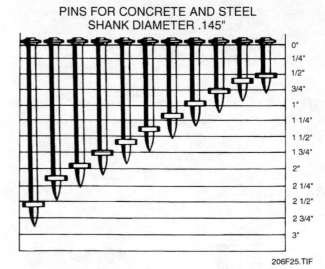

PINS FOR CONCRETE AND STEEL
SHANK DIAMETER .145"

206F25.TIF

Figure 25 ◆ Steel pins.

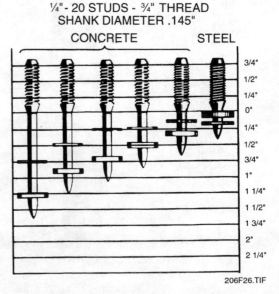

¼"- 20 STUDS - ¾" THREAD
SHANK DIAMETER .145"

CONCRETE   STEEL

206F26.TIF

Figure 26 ◆ Threaded steel studs.

6.14   PLUMBING LEVEL TWO — TRAINEE MODULE 02206

To operate a powder-actuated fastening tool (see *Figure 27*), follow these steps:

Step 1 Feed the pin or stud into the piston.

Step 2 Feed the powder booster into position.

Step 3 Position the item to be fastened in front of the tool and press it against the mounting surface. *Note:* This pressure releases the safety lock.

Step 4 Pull the trigger handle to fire the booster charge.

The booster charge releases its pressure on the piston, and the piston forces the pin or stud into the surface and stops (see *Figure 28*).

Anchors for mounting into concrete, brick, or hollow walls are available in several styles. Use wedge anchors, sleeve anchors, stud anchors, non-drilling anchors, and self-drilling anchors to fasten the piping system to concrete (see *Figure 29*). These anchors are not recommended for use in new concrete that has not had enough time to cure.

Light-duty anchors for use in solid walls, hollow block, or drywall include polyset anchors, nylon anchors, plastic inserts, expansion bolts, and spring-type toggle bolts (see *Figure 30*). All are available in a variety of sizes and lengths. Always wear safety glasses when installing anchors. Be sure to use the right size drill bit and a drill that will meet the load demands of the job.

Use concrete inserts as upper attachments to suspend pipe from a concrete structure (see *Figure 31*). Standard universal steel inserts are fabricated from heavy-gauge steel. They are designed with one case size that you can use with all sizes of support rods up through ¾ inch. After you install the inserts and remove the knockout plate, you can insert the special nuts. You can then suspend a hanging rod from the insert.

Use beam clamps when the piping is supported from the building steel (see *Figure 32*). Attach them to the bottom flange of American Standard I-beams. All are fitted with jaws that lock in position on the beam when fully and properly adjusted. Be sure to select a beam clamp that fits the thickness of the I-beam.

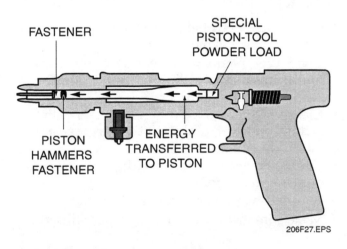

Figure 27 ◆ Powder-actuated tool.

Expanding gas from the safety booster drives a specially designed piston. The piston controls the speed and direction of the fastener, assuring safe penetration into the building material. Full penetration is reached when the piston is stopped inside the tool.

Figure 28 ◆ How the powder-actuated tool works.

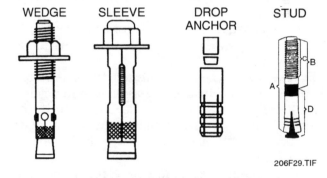

Figure 29 ◆ Anchors used for concrete or brick.

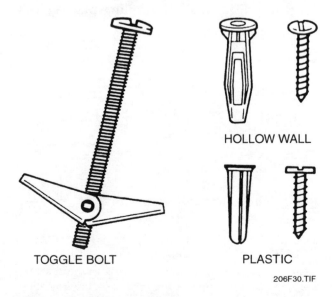

Figure 30 ◆ Light-duty anchors.

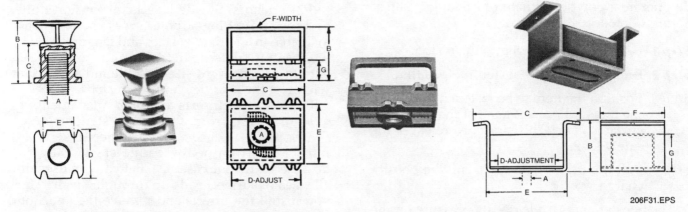

Figure 31 ◆ Concrete inserts.

Use C-clamps in installations where the pipe support is attached to I-beams, channels, or wide flange beams when it is desirable to have the support rod offset from the beam (refer to *Figure 32*). Secure the clamp to the flange with a cup-pointed, hardened setscrew. After the clamp is in place, thread a support rod into the tapped hole at the base. You can install retaining straps with the clamp to prevent the C-clamp from moving.

Use beam attachments to attach support rods to a structural member. You can weld or bolt them in either an upright or an inverted position. When you install them in an inverted position, you can make small vertical adjustments using the hanger rod.

Brackets and clips that attach threaded connections to wood beams are available in several styles. Secure them in place using bolts or screws. You may also install them on concrete structures by either bolting or welding them to the steel beam.

Ceiling flanges and plates are recommended for suspending pipelines from wood beams or ceilings (see *Figure 33*). The ceiling plate gives a

finished appearance where the rods and pipe enter the ceiling.

Use structural attachments such as washer plates, lug plates, or clevis plates to suspend support rods (see *Figure 34*). The heavy-duty washer plate is used on top of channels or angles to support the pipe with rods or U-bolts. When suspending support rods from concrete ceilings, use lug plates.

When using structural attachments, you should do the following:

- Check the specs (specifications) to see what type of hanger is required.
- Check pre-/post-stressed concrete T-sections to make sure that installing the anchors will not weaken the structure.
- Clear all locations for concrete inserts with the architect or structural engineer in charge of the building.
- Check the specs before welding attachments onto the structure.
- Avoid using powder anchors.
- Make sure that the anchor is properly inserted.

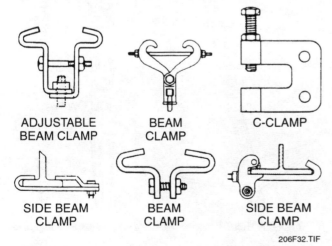

Figure 32 ◆ Beam clamps for supporting piping from the building steel.

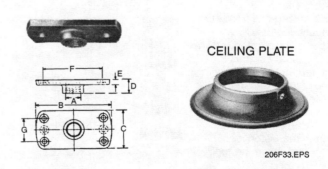

CEILING PLATE

Figure 33 ◆ Ceiling flange and plate for suspending pipelines from wood beams or ceilings.

CONCRETE CLEVIS PLATE

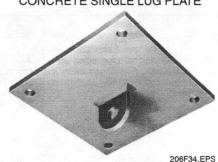
STEEL WASHER PLATE

CONCRETE SINGLE LUG PLATE

206F34.EPS

Figure 34 ◆ Structural attachments for suspending support rods.

6.3.0 Special Hangers and Accessories

Use pipe rollers to support piping that is subject to expansion or sideways movement because of temperature changes (see *Figure 35*). These rollers allow the pipes to expand and to move laterally. You can also use rollers as a rolling guide to feed the piping into place during pipe system assembly.

Use field-made shoe guides, which allow the pipe to expand and contract, to assist in pipe assembly. Two types of shoe guides you can make from materials readily available on the job site are shown in *Figure 36*. The three parts of this system are the pipe shoe, the shoe guide, and the pipe sleeve.

Install spring hangers at hanger points where vertical thermal movement occurs (see *Figure 37*). They are available in several styles and weights. Light-duty spring hangers are designed to provide a flexible spring support for light-duty loads where the vertical movement does not exceed 1¼ inches. In installations where you want to prevent vibration noise and other sounds from being transmitted into the building, use vibration control hangers to suspend the piping.

Spring cushion hangers are designed for use with a single pipe run in installations where the vertical movement does not exceed 1¼ inches. Use constant support hangers where constant, accurate support is needed on piping systems that move vertically because of temperature changes. It is a good practice to first install the pipe with rigid hangers. After the pipe is installed, replace the rigid hangers with the proper spring hangers.

Devices that protect the pipe in the hangers include protection saddles and insulation protection shields (see *Figure 38*). Protection saddles are designed for use on high temperature lines or where heat losses are to be kept at a minimum. They are also designed to transmit the pipe load to the supporting unit without damaging the covering. Insulation protection shields are recommended when low compression strength and vapor barrier type insulation such as foam or fiberglass is installed. The protection shield prevents the hanger unit from cutting, crushing, or damaging the insulation or the vapor barrier.

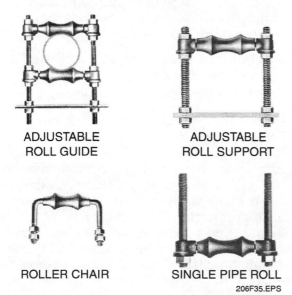
ADJUSTABLE ROLL GUIDE   ADJUSTABLE ROLL SUPPORT

ROLLER CHAIR   SINGLE PIPE ROLL
206F35.EPS

Figure 35 ◆ Pipe rollers.

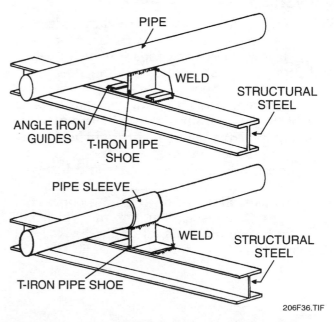
PIPE
WELD
STRUCTURAL STEEL
ANGLE IRON GUIDES
T-IRON PIPE SHOE
PIPE SLEEVE
WELD
STRUCTURAL STEEL
T-IRON PIPE SHOE
206F36.TIF

Figure 36 ◆ Field-made pipe alignment guides.

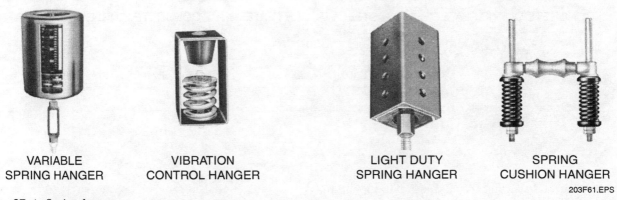

VARIABLE SPRING HANGER   VIBRATION CONTROL HANGER   LIGHT DUTY SPRING HANGER   SPRING CUSHION HANGER

203F61.EPS

Figure 37 ◆ Spring hangers.

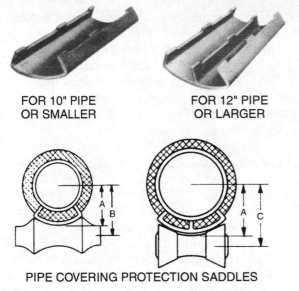

FOR 10" PIPE OR SMALLER   FOR 12" PIPE OR LARGER

PIPE COVERING PROTECTION SADDLES   INSULATION SHIELDS

206F38.EPS

Figure 38 ◆ Protection saddles and insulation shields.

6.4.0 Pipe Hanger Locations

Pipe hanger locations are governed by code. Factors governing pipe hanger locations are pipe size, piping layout, concentrated loads of heavy valves and fittings, and structural building steel available for piping support.

Support concentrated piping loads as close as possible to the load. Support terminal points close to the equipment. Locate hangers right next to any change of piping direction.

When installing pipe supports and hangers, you should be familiar with the engineer's specifications as well as local plumbing codes. For sprinkler system installation, you must be familiar with the specifications set up by the National Fire Protection Association (NFPA).

6.5.0 Supporting Vertical Piping

Support vertical piping at sufficient intervals to keep the pipe in alignment. Methods for supporting vertical piping depend on three factors:

- Whether the pipe is to be supported at or between the floor line
- The size of the pipe being supported
- Local code requirements

You can support vertical pipe at each floor line with riser clamps (refer to *Figure 17*). You may also need to install hangers that are attached between floors to the walls or vertical structural members. These hangers will maintain alignment and also will support part of the vertical load of the pipe. You can support these vertical runs between floor levels by using either an extension ring hanger and wall plate (see *Figure 39*) or a one-hole strap (see *Figure 40*).

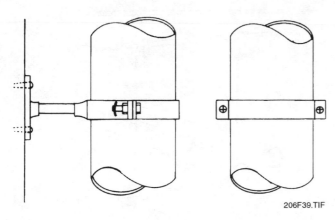

Figure 39 ◆ Extension ring hanger.

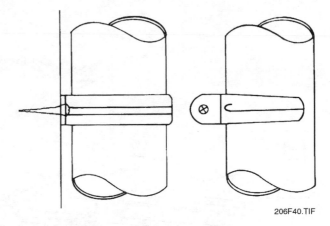

Figure 40 ◆ One-hole strap.

6.6.0 Supporting Horizontal Piping

You must support horizontal or sloping pipe at intervals that are close enough to prevent sagging and to keep the pipe in alignment. Sagging pipes produce traps that allow deposits to accumulate in low spots. Traps also can allow air or vapor to accumulate in high spots. As a general rule, each length of pipe should be independently supported so that it does not have to depend on the neighboring pipe for support. Calculate the slope on which supports are placed and the distance between supports so that each point of support is lower than the nearest upstream point (see *Figure 41*).

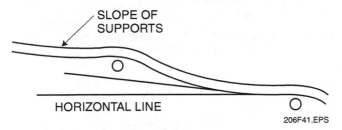

Figure 41 ◆ Slope supports.

6.7.0 Supporting Multiple Side-by-Side Runs of Pipe

In installations where multiple pipelines are run side-by-side, you can use various methods to support the pipe in one unit. Several ways that you can support multiple horizontal or vertical pipe runs using either a frame or a trapeze assembly are shown in *Figures 42, 43, 44,* and *45*. These assemblies make it possible to hang, attach, mount, frame, and support the piping system in one unit.

When you construct the trapeze or frame using channels, you create a pipe support system that you can modify by adding or removing pipe without disturbing the previously installed pipe. Sev-

eral ways in which you can use strut clips and channels when hanging and supporting side-by-side runs of pipe are shown in *Figure 46*. You can also support side-by-side runs of pipe directly on a trapeze assembly, not made from channels, using hold-down clips (see *Figure 47*).

7.0.0 ◆ MODIFYING THE STRUCTURAL MEMBERS

Occasionally, structural members lie in the path of a run of pipe. Sometimes, the space available is not sufficient to allow pipe to be run. In such cases, plumbers must modify structural members. This section deals with the five most common methods of modifying structural members:

- Drilling
- Notching
- Boxing floor joists
- Furring strips
- Building a chase

When modifying structural members, you must ensure that the structure is not weakened. If questions about the safety and strength of the structure arise, you should consult the local code or the local plumbing inspector.

7.1.0 Drilling

You may drill floor joists to allow passage for a run of pipe. A hole no larger than one-third the width of the floor joist may be drilled in its center section (see *Figure 48*). Thus, the largest hole that can be drilled in the center of a 2 × 10 floor joist is $3\frac{5}{16}$ inches in diameter. Furthermore, you must drill the hole in the middle of the joist, rather than through the upper or lower third of the joist. Only one such hole is permissible. Drilling several adjacent floor joists significantly weakens the structure.

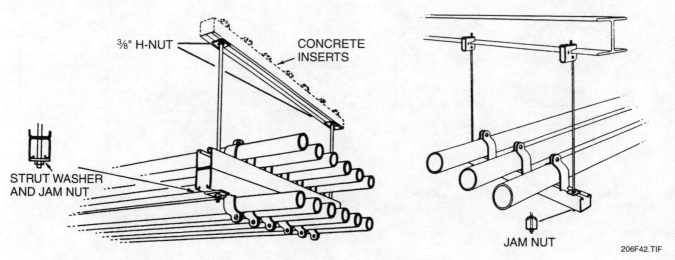

3/8" H-NUT

CONCRETE INSERTS

STRUT WASHER AND JAM NUT

JAM NUT

206F42.TIF

Figure 42 ◆ Typical applications using trapeze hangers made from one or more channel assemblies.

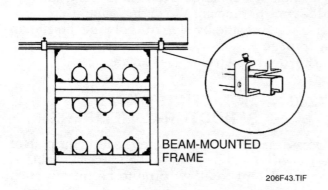

BEAM-MOUNTED FRAME

206F43.TIF

Figure 43 ◆ Beam-mounted channel frame using beam clamp.

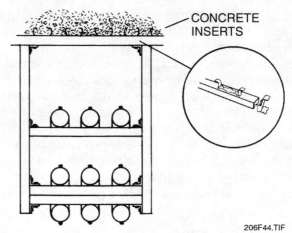

CONCRETE INSERTS

206F44.TIF

Figure 44 ◆ Ceiling-mounted channel frame using concrete inserts.

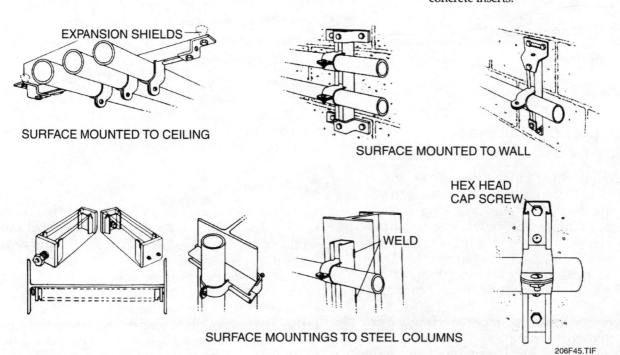

EXPANSION SHIELDS

SURFACE MOUNTED TO CEILING

SURFACE MOUNTED TO WALL

HEX HEAD CAP SCREW

WELD

SURFACE MOUNTINGS TO STEEL COLUMNS

206F45.TIF

Figure 45 ◆ Surface-mounted channel frames.

The maximum size hole that may be drilled in the end sections of the joists is 2 inches in diameter. The hole must be at least 2 inches from the edges of the board. No holes may be drilled 6 inches from either end of the floor joist.

Note that the holes drilled in the floor joists must be larger than the pipe that passes through

them to provide clearance for the pipe. Generally, you should drill a 3-inch hole for a 2-inch pipe, and a 2-inch hole for a 1½-inch pipe.

7.2.0 Notching

You may also notch structural members to provide clearance for plumbing pipes and fittings. As with drilling, local codes determine the amount and type of notching permissible. Because notching presents a greater threat to a building's structural strength than drilling, notching should be done only when absolutely necessary.

Notching is restricted in the center one-third of structural members (see *Figure 49*). Therefore,

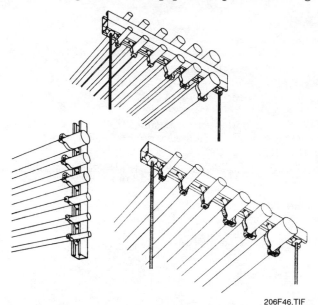

206F46.TIF

Figure 46 ◆ Side-by-side runs of piping supported using strut clips and channels.

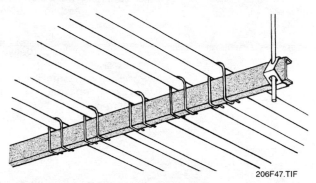

206F47.TIF

Figure 47 ◆ Supporting piping directly on steel or wood trapeze using multipurpose clips.

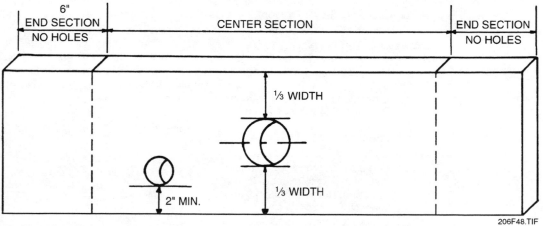

206F48.TIF

Figure 48 ◆ Drilling floor joists.

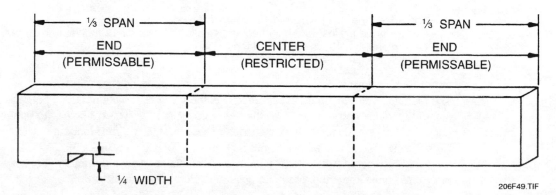

206F49.TIF

Figure 49 ◆ Notching structural members.

Plumbing Code Requirements

Plumbing codes regulate the maximum interval allowed between supports. Plumbing codes vary from area to area, but each may be based on one of several model codes. Check your local code. Following is a sample of model code requirements for hanger and support locations as they apply to cast-iron pipe, threaded pipe, copper tubing, lead pipe, and plastic pipe.

The Uniform Plumbing Code (UPC)

Vertical Piping
- Cast-iron pipe: Every story or closer
- Threaded pipe: Not less than every other story
- Copper tubing: Each story or at maximum intervals of 10 feet
- Lead pipe: Intervals not exceeding 4 feet
- Plastic pipe: Not addressed

Horizontal Piping
- Cast-iron pipe: Not more than 5-foot intervals where joints occur
- Pipe exceeding 5 feet in length may be supported at intervals not exceeding 10 feet
- Supports must be within 18 inches of the hub or joint
- For hubless or compression-gasketed joints, support must be at least every other joint unless the length between supports exceeds 4 feet—then support must be provided at each joint.
- Each horizontal branch connection must be supported
- Threaded pipe: 10-foot intervals for ¾ inch and smaller; 12-foot intervals for 1 inch and larger
- Copper tubing: 6-foot intervals for 1½ inches and smaller; 10-foot intervals for 2 inches and larger
- Lead pipe: Continuous support along entire length
- Plastic pipe: Not to exceed 4 feet

The Standard Plumbing Code (SPC)

Vertical Piping
- Cast-iron pipe: Every story level and at intervals not to exceed 15 feet
- Threaded pipe: Not less than every other story at intervals not to exceed 30 feet
- Copper tubing: Each story for 1½ inches and over; not more than 4-foot intervals for 1¼ inches and smaller

Horizontal Piping
- Cast-iron pipe: 5-foot intervals on 5-foot lengths; 10-foot intervals on 10-foot lengths
- Threaded pipe: Not to exceed 12 feet
- Copper tubing: 6 feet for 1½ inches and smaller; 10 feet for 2 inches and larger
- Lead pipe: Continuous support along entire length
- Plastic pipe: Not to exceed 4 feet

National Plumbing Code

Vertical Piping
- Cast-iron pipe: Not to exceed 15 feet
- Threaded pipe: Not to exceed 15 feet
- Copper tubing: Not to exceed 10 feet
- Lead pipe: Not to exceed 4 feet
- Plastic pipe: Not to exceed 4 feet

Horizontal Piping
- Cast-iron pipe: Not to exceed 5 feet
- Threaded pipe: Not to exceed 12 feet
- Copper tubing: 6 feet for 1¼ inches and smaller; 10 feet for 1½ inches and larger
- Lead pipe: Continuous support along entire length
- Plastic pipe: Not to exceed 4 feet

Other Requirements

The Manufacturers Standardization Society of the Valve and Fitting Industry publishes a standard that is often referred to by the model building codes when it comes to hangers and supports. This standard, which is the industry source, is titled *Pipe Hangers and Supports: Selection and Application* and is available from the Society as SP-69-83.

In addition, some of the model plumbing codes also contain information about specialty piping. You must know and follow the governing plumbing code.

notching can safely be done only near the ends of the floor joists. Also, the depth of the notch cannot exceed one-fourth of the width of the board.

In cases where you are forced to notch structural members, you should use bracing. In most cases, the braces consist of short lengths of 2 × 4s nailed to both sides of the notched member (see *Figure 50*). Bracing significantly strengthens the member.

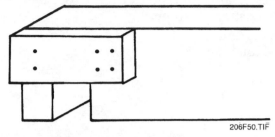

Figure 50 ◆ Bracing structural members.

7.3.0 Boxing Floor Joists

Where either drilling or notching will not provide adequate plumbing clearance, plumbers use the boxing method. Boxing floor joists is commonly necessary where a floor joist interferes with the installation of the closet flange (see *Figure 51*). In this case, cut the end of the floor joist back to clear the closet bend and flange. Then construct a double header to bridge the gap between the two adjacent floor joists and nail it into place.

Strap hangers are another method of fastening the floor joists to the header (see *Figure 52*). To further strengthen the floor, double the adjacent joists.

7.4.0 Furring Strips

The space inside the framed wall must be wide enough to allow passage of the water supply piping. If it is not, add furring strips (see *Figure 53*). When you add furring strips, remember that you must locate the fixtures from the furred wall and not from the original wall.

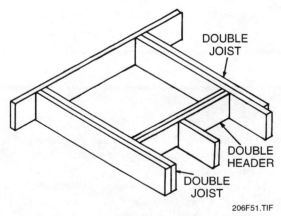

Figure 51 ◆ Boxing floor joists.

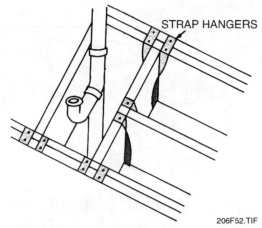

Figure 52 ◆ Using strap hangers.

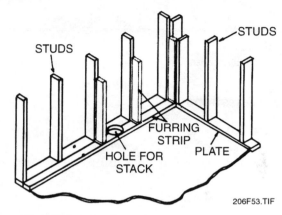

Figure 53 ◆ Using furring strips.

7.5.0 Building a Chase

In some instances, even furring strips will not provide the necessary clearance for plumbing pipes and fixtures. In these situations, build a chase (see *Figure 54*). The chase is an extremely small, sealed space that provides the necessary clearance for pipes, fittings, carriers, and other pieces of plumbing. The chase may or may not extend to the ceiling and adjacent walls. Usually, it is permanently sealed, so you must properly install the plumbing before the wall-finishing material is applied. Make cleanouts accessible with removable cover plates flush-mounted on the chase wall.

WARNING!

Be sure to check the local code or consult with the plumbing inspector before modifying any structural members. Improper modification can seriously weaken a building's support structure.

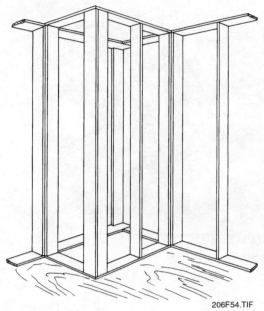

Figure 54 ◆ The chase.

206F54.TIF

Review Questions

Sections 5.0.0–7.5.0

1. Install the water supply piping in reference to the _____.
   a. fixture rough-ins
   b. walls
   c. drains
   d. finished wall treatment

2. Vertical pipes should be supported at each floor using _____.
   a. riser clamps
   b. connectors
   c. universal pipe clamps
   d. perforated band irons

3. Horizontal pipe supports are necessary to prevent sagging and keep pipe _____.
   a. in side-by-side runs
   b. properly vented
   c. well insulated
   d. in alignment

4. A hole drilled in a floor joist to allow for a run of pipe must be _____.
   a. no more than one-third the width of the joist
   b. within four inches from the end of the joist
   c. at least one inch from the edge of the joist
   d. at least one inch larger than the pipe

5. _____ is recommended when a floor joist interferes with installation of the closet flange in water supply piping.
   a. Furring
   b. Notching
   c. Drilling
   d. Boxing

DID YOU KNOW?

Herod's Human Water Supply Line

In 38 B.C.E., Herod ruled Judea. He left his mark on the land, most notably on Masada, a 1,300-foot-high rock fortress in the middle of a desert east of the Dead Sea. Its water system originated in two small *wadis,* or gullies. These wadis quickly flooded with water from sudden, unpredictable downpours. To hold the water until needed, the Romans constructed dams in two places. On demand, the dams allowed the water to flow by gravity through an aqueduct and channels directed to the site. In addition, a set of cisterns at the top of Masada connected to conduits to catch rainwater. But the water supply was not enough for the palaces located there. So Herod ordered a human conduit to bring up water from the cisterns far below. Some historians estimate that hundreds, maybe thousands, of slaves and beasts of burden carried jars of water up the cliff for the royal households, for drinking, cooking, and luxurious steam baths. Today Masada is a popular tourist attraction and an Israeli national shrine.

ON THE · LEVEL ·

Clean Pipes

You must make every effort to keep the water supply piping as clean as possible before installation. Sand, gravel, or mud trapped in the pipes will flow through the piping system, lodge in the valves and faucets, and damage the washers and O-rings. The faucets will then leak. Store pipe in a clean, dry area and cap all ends of pipe at the end of each workday. Be sure to flush the line as thoroughly as possible to remove any sand, mud, or gravel from the system.

8.0.0 ◆ MAIN SUPPLY LINES

Once you locate the water heater, water softener (if required), and hose bibbs, and assemble and install the fixture riser and stubouts, you can install the main supply lines.

For residential installations, make sure that the main feeder line beyond the water heater is the right size to supply the required flow and pressure. Using smaller pipe sizes can save both energy and money.

For example, a ¾-inch diameter pipe has a cross-sectional area of .4418 square inches, and a ½-inch diameter pipe has a cross-sectional area of .1963 square inches (see *Figure 55*). This means that a ¾-inch pipe will hold 2¼ times the amount of water that a ½-inch pipe will hold. With a larger pipe, a faucet will have to run longer to get hot water to a sink. In addition, the amount of hot water that will be left in the pipe to cool once the faucet is turned off is significantly greater with larger pipe. Choosing the smaller pipe can result in potential energy savings for the customer. Always consult the local codes to determine the water- and energy-saving measures possible for each job. Note that on larger residential and commercial applications, plumbers install a recirculating pipe to return the water in the supply line to the water heater.

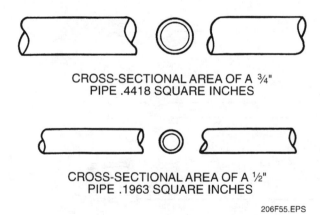

CROSS-SECTIONAL AREA OF A ¾"
PIPE .4418 SQUARE INCHES

CROSS-SECTIONAL AREA OF A ½"
PIPE .1963 SQUARE INCHES

206F55.EPS

Figure 55 ◆ Cross-sectional areas of pipe.

In multifamily units, control valves are required to isolate and repair each living unit. As the branch runs are installed, you should cut, prepare, and join several lengths of pipe and fittings before soldering. This saves time and provides for a smoother installation.

When you install the cold water lines, it is important to consider the possible branching off of the line to the hose bibbs to save pipe and time. If the cold water lines are to carry softened water, this will not be possible. It is not economically feasible to soften the cold water delivered to the hose bibbs for lawn sprinkling or other outdoor applications.

8.1.0 Water Hammer

When liquid flowing through a pipe is suddenly stopped, vibration and pounding noises—called *water hammer*—result. The forces generated at the point where the liquid stops are like an explosion. When water hammer occurs, a high-intensity shock wave travels back through the pipe until it reaches a point of relief. That point might be a large diameter riser or piping main (see *Figure 56*). The shock wave pounds back and forth between the point of impact and the point of relief until the destructive energy dissipates.

The quick closing of electrical, pneumatic, or spring-loaded valves as well as the quick hand-closing of valves or fixture trim is the cause of water hammer. Note that noise does not always occur with water hammer. Quick closing of valves always creates some degree of shock—with or without noise.

Water hammer arresters prolong the service life of the components that make up a water supply system (see *Figure 57*).

8.2.0 Air Chamber

The air chamber also has been used as a way of controlling shock. The unit consists of a capped piece of pipe that is the same diameter as the line it serves. Its length ranges from 12 inches to 24 inches. Air chambers may be constructed in several different shapes (see *Figure 58*).

8.3.0 Other Water Supply Connections

In both residential and commercial installations, you may have to provide piping for miscellaneous appliances such as dishwashers, washing machines, and icemakers.

Dishwashers may require a tee or a three-way water-stop off the hot water access line for the kitchen sink. You won't need a separate drain for the dishwasher because its wastewater is pumped through the sink drain above the trap or into the garbage disposal unit.

Washing machines require both hot and cold water lines. They also require a hose bibb for each line so that the hoses from the washer can be connected to the lines. Locate these faucets so that the customer will have easy access after the washer is installed.

Refrigerators with icemakers require both a line to bring in water and an accessible stop valve to turn the water supply on or off. Sometimes the valve is located beneath the sink and sometimes it is located in a box behind the refrigerator (see *Figure 59*).

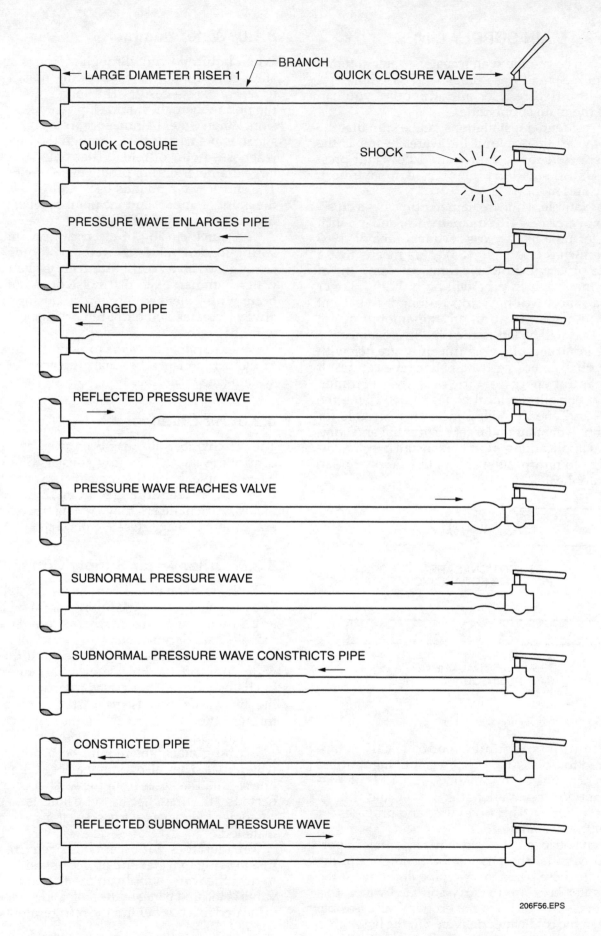

LARGE DIAMETER RISER 1 — BRANCH — QUICK CLOSURE VALVE

QUICK CLOSURE — SHOCK

PRESSURE WAVE ENLARGES PIPE

ENLARGED PIPE

REFLECTED PRESSURE WAVE

PRESSURE WAVE REACHES VALVE

SUBNORMAL PRESSURE WAVE

SUBNORMAL PRESSURE WAVE CONSTRICTS PIPE

CONSTRICTED PIPE

REFLECTED SUBNORMAL PRESSURE WAVE

206F56.EPS

Figure 56 ◆ Shock wave.

206F57.EPS

Figure 57 ◆ Water hammer arrester.

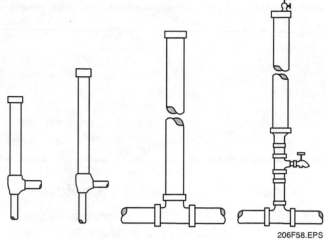

206F58.EPS

Figure 58 ◆ Air chamber.

9.0.0 ◆ COMPLETING THE INSTALLATION

As you work on a water supply system installation, you need to anticipate potential problems and provide ways to either prevent them or allow for quick repair. Among the things to keep in mind are accessibility for service repair, protection from frost, protection from punctures, and backflow prevention.

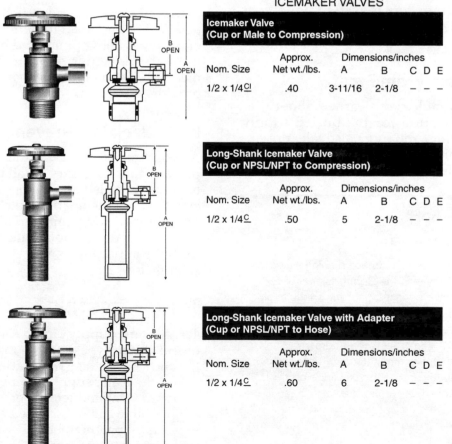

ICEMAKER VALVES

Icemaker Valve (Cup or Male to Compression)

Nom. Size	Approx. Net wt./lbs.	Dimensions/inches				
		A	B	C	D	E
1/2 x 1/4 CI	.40	3-11/16	2-1/8	–	–	–

Long-Shank Icemaker Valve (Cup or NPSL/NPT to Compression)

Nom. Size	Approx. Net wt./lbs.	Dimensions/inches				
		A	B	C	D	E
1/2 x 1/4 C	.50	5	2-1/8	–	–	–

Long-Shank Icemaker Valve with Adapter (Cup or NPSL/NPT to Hose)

Nom. Size	Approx. Net wt./lbs.	Dimensions/inches				
		A	B	C	D	E
1/2 x 1/4 C	.60	6	2-1/8	–	–	–

206F59.EPS

Figure 59 ◆ Icemaker valves.

Air Chambers

Air chambers are more economical than manufactured water hammer arresters. What happens with air chambers? The air that acts as a cushion between the water and the air chamber cups will eventually become saturated with water. When that happens, the air chamber is no longer effective. To remedy this, you can turn off the water to the building and drain the water pipe. Once the pipe is drained, turn the water supply on. The air chamber will recharge with air. Repeat as necessary.

9.1.0 Accessibility for Service Repair

Sometimes shut-off valves are located in areas that are not easily accessible once the building is completed. For example, the shut-off valves to a tub or shower are located in the finished wall. Without some form of easy access, it will be necessary to cut a hole in the wall to get to the valves. Generally plumbers will provide an access panel—a panel that is easily unscrewed and removed—to allow easy access to the shut-off valve. When doing installations you should try to anticipate the need for access to the valves and provide it if possible.

In basements or crawl spaces, access to cleanouts as well as to shutoff valves and water meters is necessary. If the structure is equipped with a gravity drain-down system, you must install strategically located valves at the low points of each line. These valves also require easy access.

9.2.0 Protection From Frost

Earlier in this module, you learned about providing frost protection for the buried supply lines to the building. You must also take precautions for frost protection inside the building. Avoid running plumbing on exterior walls, espe-

cially in colder areas. When you must locate sinks or other fixtures on exterior walls, run the supply piping up through the floor so that the pipes are inside the building. If the local code permits it, insulate the water supply piping inside walls.

Crawl spaces can also cause problems for water supply piping because they are often vented to the exterior. In many cases, crawl space vents are closed for the winter season, which should protect the plumbing. If water supply piping passes in front of or close to vents, the potential for freezing may increase. Care in pipe location is necessary for a trouble-free installation.

9.3.0 Protection From Punctures

During and after construction, pipes are vulnerable to punctures. To minimize the possibility of the pipes being punctured by a nail, run pipes through the center of a 2 × 4 wall. Nail sheet metal plates to either side of the 2 × 4 studs to provide protection.

9.4.0 Backflow Prevention

As you learned earlier in this module, backflow prevention is necessary on all installations where hoses are installed and the possibility of backflow exists. These installations include dishwashers, hose bibbs, and slop sinks. Although some plumbing codes do not require **vacuum breakers** for these applications, you should seriously consider their use.

10.0.0 ◆ TESTING

Test all water supply piping to ensure that it is free of leaks. Two types of tests may be conducted—the **air test** and the **hydrostatic** (water) **test.**

Test the water supply piping system before it is enclosed. On smaller systems, all of the water supply piping may be tested at one time. Larger projects—apartment buildings, for example—may require that you test the water supply piping in sections. Sometimes this is done floor by floor.

 DID YOU KNOW?

A Change in Plans

The White House did not have running water until 1831. That year, the Commissioner of Public Buildings bought a bubbling spring to pipe water to the White House in wood pipes made from drilled-out logs. The piping might have been installed earlier, in 1829 when Andrew Jackson took office, but the Committee on Public Buildings decided to improve the building's north entrance instead. The delay may have been beneficial. As the project got under way, the engineer exchanged the wooden pipes for pipes made of iron.

Typically, plumbers will conduct at least one test before scheduling the test with the inspector. This provides an opportunity to make any necessary repairs before the official test. The plumbing contractor is responsible for furnishing all materials, equipment, and labor to conduct the test. The plumbing inspector observes the test and verifies that it is successful.

10.1.0 The Air Test

Air tests for water supply piping are similar to the air tests for DWV piping. It's a fairly simple procedure that requires a limited number of tools.

A **test gauge assembly** enables compressed air to enter the piping system (see *Figure 60*). It also provides a gauge for measuring the air pressure within the piping system.

You can use a **compressor** to produce the required pressure (see *Figure 61*). A large air compressor may be on the construction site to operate a variety of air tools. If a large compressor is not available, use a small electrical or gasoline-powered unit. In either case, the compressor should provide air at a pressure above 150 pounds per square inch (psi).

Plumbers don't generally use test caps, or **test plugs** to close openings in the water supply piping system during testing. However, they are available and may be useful in selected applications. These caps are removed at the time of fixture installation and are not recommended for regular service use. The more common method of closing openings is to install regular service caps or plugs.

> **CAUTION**
>
> Do not do air tests on PVC piping. The pressure can shatter the pipe. Check with the manufacturer or local code before doing an air test on PVC water piping.

10.2.0 Air Test Procedures

When conducting an air test, follow these steps:

Step 1 First inspect the system for defects and missing parts. During this inspection, you must close all openings with caps, plugs, test plugs, or valves.

Step 2 Attach the test gauge assembly. Make certain that it is securely joined to the piping system. Use pipe tape or **pipe dope** on threaded connections.

Step 3 Open the valve and introduce air until the required pressure is recorded on the gauge. The exact pressure required will be dictated by local code—for example, 1½ times the normal water pressure. Make sure that the valve on the test gauge assembly is closed. This prevents leaks during testing. The test is considered successful if the water supply piping will hold the required pressure for a specified period, which may be up to 24 hours.

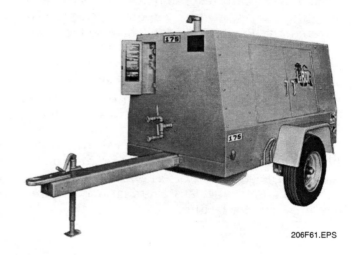

206F61.EPS

Figure 61 ◆ Air compressor.

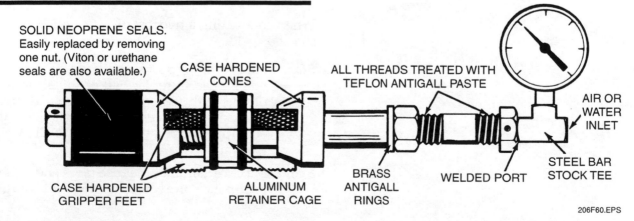

206F60.EPS

Figure 60 ◆ Test gauge assembly.

Step 4 If the piping fails the test, you must locate the defective or leaking joint. A soap solution may be helpful in identifying small leaks. Spray or spread the solution over the area and watch for bubbles to form to locate the leak. Once you have repaired the leak, you must test the system again.

CAUTION

Test caps, or plugs, are not recommended for air tests in copper pipe. The air pressure might rupture the cap.

10.3.0 The Hydrostatic Test

Hydrostatic (water) tests are conducted by filling the pipe with water. Generally, this process is used for larger-diameter piping such as water mains.

Use a **hydrostatic test pump** to conduct the test. The pump may be manually operated, electrically powered, as shown in *Figure 62*, or gasoline powered, as shown in *Figure 63*.

Powered pumps are rated by the number of gallons per minute (gpm) they put out and the maximum pressure (in psi) they produce. It is important to size the test pump to the requirements of the system being tested.

High-pressure hydrostatic testers also are available (see *Figure 64*). These units can produce up to 3,000 psi. They are used to test boilers, tanks, heat exchangers, and other systems that may require high pressures for testing.

206F63.EPS

Figure 63 ◆ Gas-powered hydrostatic test pump.

206F64.EPS

Figure 64 ◆ High-pressure hydrostatic tester.

10.4.0 Hydrostatic Test Procedures

When conducting a hydrostatic test, follow these steps:

Step 1 Fill the piping system with water. Because you'll need to replace all of the air in the piping with water, install valves at high points in the system where air is likely to be trapped. Open these valves to allow air to escape.

Step 2 Make certain that the discharge end of the pump is connected to the test gauge assembly.

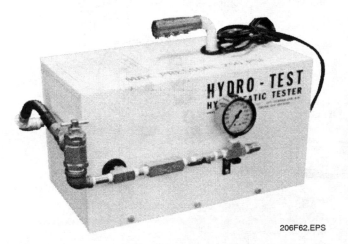
206F62.EPS

Figure 62 ◆ Electrically powered hydrostatic test pump.

Step 3 Fit a hose onto the inlet (suction) end of the pump and connect it to a water supply or insert it into a bucket of water to provide extra water for compression. Water compresses very little; therefore, you need to force only a small amount of additional water into the piping to achieve the required pressure. If the pressure gauge fluctuates, or if you cannot obtain the required pressure, either a leak is present or air remains in the piping. You can easily detect a leak by observing where the water escapes. If trapped air is the problem, open the valves and allow the air to escape.

WARNING!
Never immerse an electrical test unit in water. You could be electrocuted.

CAUTION
Never leave a hydrostatic test pump unattended. The pressure can build quickly and may damage the system.

Step 4 Maintain the proper pressure and time for the test. The local code or job specifications will indicate the appropriate pressure to maintain and the length of the test.

Step 5 Slowly open a valve to relieve the pressure once the test is concluded.

Step 6 Drain all water to prevent damage to the pipe from freezing.

10.5.0 Test Pump Operation

The following tips apply to using both electrically powered and gasoline-powered hydrostatic testing pumps:

- Always use clean, cold water.
- Use water pressure feed when operating electric test pumps. *Never* leave a test pump unattended. The inlet pressure may vary from 5 psi to 100 psi. Gasoline test pumps, on the other hand, are designed to work from gravity feed or siphon from a barrel or tank. If required, gasoline-powered units may also work from water pressure feed.

- Be sure the water flow is on before operating the unit. *Never* run a test pump without water for more than 30 to 45 seconds. Severe damage to the machine may result.
- Manufacturers of testing equipment suggest that you install a tee and a valve on the running side of the tee at the system being tested and use it for a bleed-down valve.
- Be sure you know what pressure the relief valve is set for before you begin the test. The setting of the relief valve determines the maximum pressure that the test pump can generate.
- Be sure to supply an electrical test unit with a grounded source of electricity. The power source must match the needs of the motor.
- Most test pumps are designed for use with water only. Avoid using other fluids.
- The time required to test a system depends on the amount of water needed to pressurize the system, not necessarily the size of the system. The incoming water displaces the air within the system and creates pressure. Therefore, the more air that is trapped within the system before testing, the longer it takes to pressurize the system and conduct the test.
- If it seems that it is taking a longer time than normal to test a system, look for a leak, an open valve, or some other way that the system might be losing water. If you suspect that the hydrostatic tester itself may not be working, test it by putting a valve on the outlet end of the outlet hose. Then, with the hydrostatic tester running, slowly turn the valve off and check the pressure gauge. If the gauge registers pressure, the unit is working.

Review Questions

Sections 8.0.0–10.5.0

1. Using _____ pipes for the main water supply line saves both energy and money for a single family residence.

   a. recirculating
   b. higher volume
   c. smaller diameter
   d. larger diameter

2. An air chamber is an economical way to control water hammer when _____.

   a. water hammer occurs only in cold weather
   b. all water in the supply system is softened
   c. water hammer occurs only in warm weather
   d. the air chamber water pipe is recharged with air periodically

3. The air test on a water supply system measures air pressure for _____.

   a. up to 24 hours
   b. up to 4 hours
   c. at least 6 hours
   d. no more than 1 hour

4. To prevent backflow, install a(n) _____.

   a. union
   b. air chamber
   c. vacuum breaker
   d. universal pipe clamp

5. A hydrostatic test of a water supply system should occur with _____.

   a. air still in the system
   b. valves installed to remove air from the system
   c. a compressor to put water under substantial pressure
   d. valves open to completely flush the system of water

Summary

Installing water supply piping is similar to installing DWV piping. You develop a material takeoff, locate fixtures, and determine the route of the piping. You'll base your design for the water supply system on plans and specifications. However, you'll have some flexibility to install the system in a cost-effective way. And you can modify the plans to solve problems on site. Some of the items to consider are the demand for water, whether a water softener is included, and the number of hose bibbs. In addition, the job may include fixtures like dishwashers, washing machines, and icemakers.

It is important to hang and support the pipes securely, making sure that you don't weaken any structural members that you modify. You'll also need to protect the pipes from nail punctures, freezing, and water hammer.

An important part of the installation is to conduct either an air test or a hydrostatic test to ensure that the system is leak-free and ready for the customer's use.

Trade Terms Introduced in This Module

Air test: A test to determine a pipe's ability to withstand air pressure and to detect leaks.

Backflow: The flow of water or other liquids into the distributing pipes of a potable water supply from somewhere other than its intended source.

Compressor: A machine for compressing air or other gases.

Curb box: A vertical sleeve that provides access to a buried curb stop.

Curb stop: In a water service pipe, a control valve for a building's water supply, usually placed between the sidewalk and the curb.

Dielectric fitting: In a water supply system, a type of adapter (for example, a union) that connects a copper pipe with an iron pipe. It is used to prevent galvanic action, which can cause corrosive failure in pipes.

Frost line: The depth of frost penetration into the soil. The depth varies in different parts of the country. Water supply and drainage pipes should be set below this line to prevent freezing.

Hose bibb: An exterior water faucet, usually threaded to provide a connection for a hose.

Hydrostatic test: A test to determine a pipe's ability to withstand internal hydrostatic (water) pressure and to detect possible leaks in the system.

Hydrostatic test pump: A pump used to test hydrostatic pressure.

Pipe dope: A compound used to lubricate a pipe joint and to make it leakproof.

Test gauge assembly: A device that enables compressed air to enter the piping system; it also provides a gauge for measuring the air pressure within the piping system.

Test plug: A plug that is installed in a system being tested for leaks.

Vacuum breaker: A backflow preventer that prevents a vacuum in a water supply system from causing backflow.

Answers to Review Questions

Sections 2.0.0–4.3.0	Sections 5.0.0–7.5.0	Sections 8.0.0–10.5.0
1. c	1. c	1. c
2. b	2. a	2. d
3. c	3. d	3. a
4. a	4. a	4. c
5. d	5. d	5. b

ACKNOWLEDGMENTS

Figure Credits

Plumbing and Drainage Institute   206F56, 206F58

NIBCO   206F59

Sioux Chief Manufacturing Company   206F57

NCCER CRAFT TRAINING USER UPDATES

The NCCER makes every effort to keep these textbooks up-to-date and free of technical errors. We appreciate your help in this process. If you have an idea for improving this textbook, or if you find an error, a typographical mistake, or an inaccuracy in the NCCER's Craft Training textbooks, please write us, using this form or a photocopy. Be sure to include the exact module number, page number, a detailed description, and the correction, if applicable. Your input will be brought to the attention of the Technical Review Committee. Thank you for your assistance.

Instructors – If you found that additional materials were necessary in order to teach this module effectively, please let us know so that we may include them in the Equipment and Materials list in the Instructor's Guide.

Write: Curriculum Revision and Development Department
               National Center for Construction Education and Research
               P.O. Box 141104, Gainesville, FL 32614-1104

Fax: 352-334-0932

E-mail: curriculum@nccer.org

Craft _____ Module Name _____

Copyright Date _____ Module Number _____ Page Number(s) _____

Description _____

_____

_____

_____

(Optional) Correction _____

_____

_____

(Optional) Your Name and Address _____

_____

_____

Installing Fixtures, Valves, and Faucets

COURSE MAP

This course map shows all of the modules in the second level of the Plumbing curriculum. The suggested training order begins at the bottom and proceeds up. Skill levels increase as you advance on the course map. The local Training Program Sponsor may adjust the training order.

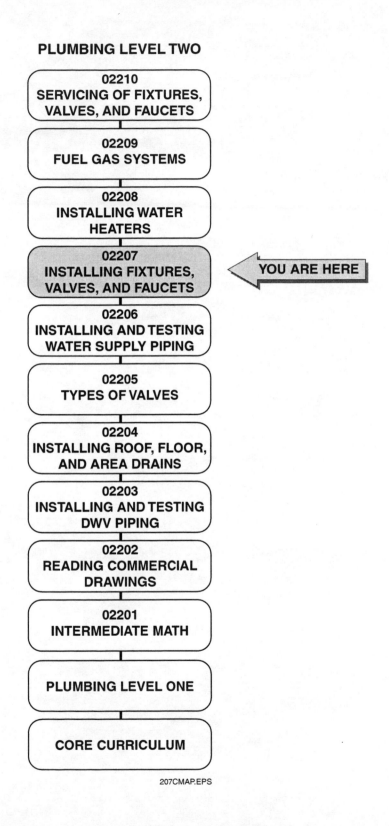

PLUMBING LEVEL TWO

02210
SERVICING OF FIXTURES, VALVES, AND FAUCETS

02209
FUEL GAS SYSTEMS

02208
INSTALLING WATER HEATERS

02207
INSTALLING FIXTURES, VALVES, AND FAUCETS

YOU ARE HERE

02206
INSTALLING AND TESTING WATER SUPPLY PIPING

02205
TYPES OF VALVES

02204
INSTALLING ROOF, FLOOR, AND AREA DRAINS

02203
INSTALLING AND TESTING DWV PIPING

02202
READING COMMERCIAL DRAWINGS

02201
INTERMEDIATE MATH

PLUMBING LEVEL ONE

CORE CURRICULUM

207CMAP.EPS

MODULE 02207 CONTENTS

Figures

Installing Fixtures, Valves, and Faucets

Objectives

When you have completed this module, you will be able to do the following:

1. Describe the general procedures you should follow before installing any fixture.
2. Demonstrate the ability to install bathtubs, shower stalls, valves, and faucets.
3. Demonstrate the ability to install water closets and urinals.
4. Demonstrate the ability to install lavatories, sinks, and pop-up drains.
5. Demonstrate how to protect fixtures.

Prerequisites

Before you begin this module, it is recommended that you successfully complete the following modules: Core Curriculum; Plumbing Level One; Plumbing Level Two, Modules 02201 through 02206.

Required Trainee Materials

1. Paper and sharpened pencil
2. Appropriate Personal Protective Equipment

1.0.0 ◆ INTRODUCTION

In this module, you'll learn how to install basic plumbing fixtures. Plumbing fixtures are the connection between the potable water supply and the sewage system. Liquid waste flows into the fixture before it flows into the drainage system. Common residential fixtures include bathtubs, showers, lavatories, sinks, water closets, and, in some cases, urinals, which are more common in commercial installations. This module includes general instructions that apply to all fixture installations. Because fixtures come in a wide variety of models, it's important to follow the manufacturer's instructions for specific information and to prevent voiding the manufacturer's warranty.

In this module, you'll also learn how to install valves and faucets. A faucet is a valve installed at the end of a water supply line.

Fixtures can be divided into two categories:

- **Floor-mounted** (bathtubs and water closets)
- **Wall-hung** (lavatories and urinals)

When installing fixtures, you must follow proper installation procedures. These procedures are based on accepted engineering practices. Remember the following:

- The little bit of extra time you take to follow correct procedures will prevent **callbacks.** Callbacks are time-consuming and costly.
- Fixture installation is called **finish work** because once the walls are closed up, it's the one part of your plumbing job that everyone—including your boss and your customer—will see. Make sure your finish work reflects your best skill and workmanship. The care and skill you apply to the installation inside the walls should continue to the finish work.

CAUTION

Changes may affect the installation of fixtures. Always check the rough dimensions to verify measurements and placement.

2.0.0 ◆ BEFORE YOU INSTALL THE FIXTURES

A wholesaler may deliver the fixtures directly to the job site, or you may pick them up yourself from a supplier or from a plumbing contractor's warehouse. After you get the fixtures, follow these procedures:

• Check the brand and model to ensure that the fixtures meet the job specifications. Also, check to see if fixture trim such as faucets, traps, and strainers are included. Make sure that each trim piece fits the fixture.

• Examine the condition of the fixtures and trim. Open all containers delivered to the job site while the delivery person is there. Doing this eliminates questions about who is at fault if damage is discovered later. If you pick up the fixtures and trim directly from a supplier or contractor's warehouse, inspect them before you leave. Check the contents of previously opened cartons for missing items.

• Verify that the **rough-in** dimensions are correct. This is a critical step. Errors could have occurred in the initial rough-in. Changes in the building, such as wall placement and wall covering, could affect your work. You might have to install a fixture that is different from the one originally specified.

• Protect the fixtures before installation. Ideally, you'll schedule delivery of the fixtures right before you are ready to install them. But if that's not possible, protect the fixtures by storing them in a clean, dry space that you can lock.

• Save all of the packing materials. You can use the package to protect the installed fixtures from weather and the work of other trades. Also, you'll need the package to return an unsuitable fixture or piece of trim.

• Save all of the manufacturer's installation and care instructions and warranty materials. You'll turn these over to the customer when the job is completed.

• Protect the installed fixtures until the building is ready for occupancy. Cut and fit cartons over the installed fixtures to protect surfaces from scratches and chips.

• Protect your customers. They expect to be the first to use their new facilities. Turn the water to the fixture off and place DO NOT USE signs on the covered fixtures. On large projects, you may install temporary fixtures in one restroom for workers to use. Replace these fixtures if necessary before you complete the project.

Follow these procedures on every installation job. You'll reduce or even eliminate mistakes, and you'll save time and money.

3.0.0 ◆ INSTALLING BATHTUBS AND SHOWER STALLS

The following installation steps are provided as a general guide. *Always* read and follow the manufacturer's instructions to determine the specific instructions for bathtubs and shower stalls.

You'll deal with bathtubs and shower stalls first. These fixtures are too large to install later. Follow these steps to install a typical bathtub:

Step 1   Check the rough-in dimensions.

Step 2   Install bracing for the tub (see *Figure 1*).

Step 3   Turn the tub on its side and install the tub waste and overflow piping according to the manufacturer's directions. Note that moving the tub may require two people.

Step 4   Move the tub into position. Make certain that it rests firmly on the bracing.

Step 5   Set the tub and shower valve and brace properly.

Step 6   Install the shower riser, elbow, shower arm, and head (see *Figure 2*).

Step 7   Install the filler drop, elbow, nipple, and spout.

Step 8   Connect the waste tailpiece drain system.

Step 9   Fill the bathtub through the valve. You should fill the tub to overflow.

Step 10   Check for leaks.

Step 11   Turn off the water and cover the tub with packing material to protect it.

ON THE  ·LEVEL·

Installations for the Physically Challenged

When doing installations for the physically challenged, you must coordinate the height of valves and faucets to grab bars.

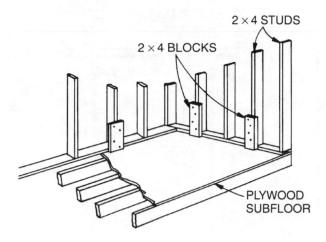

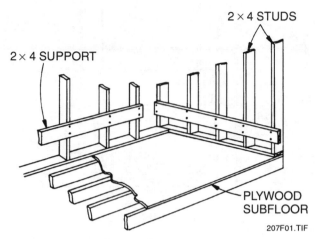

Figure 1 ◆ Bracing for the tub.

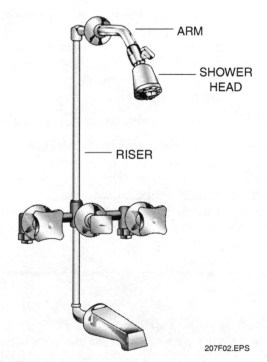

Figure 2 ◆ Shower riser, arm, and shower head.

3.1.0 Installing a Fiberglass Tub and Shower Unit

Always follow the manufacturer's instructions when positioning the required **blocking**, grout, or other supporting material (see *Figure 3*). Trial fit the unit to ensure that the bearing points are properly supported. Fiberglass has a tendency to crack under stress, so you must correctly fit the framing around units made of this material. Avoid bending the material during installation. Bending may result in cracks that will not become visible until later.

4.0.0 ◆ INSTALLING VALVES AND FAUCETS

Valves and faucets control the flow of fluids or gases through a pipe. You must install them properly to ensure that they do their job. There are five common types of valve connections, or joints:

- Threaded
- Soldered
- Solvent welded
- Flanged
- Compression

Determine the type of connection to use by considering the plumbing application, ease of installation, and the material that makes up the pipe and valves.

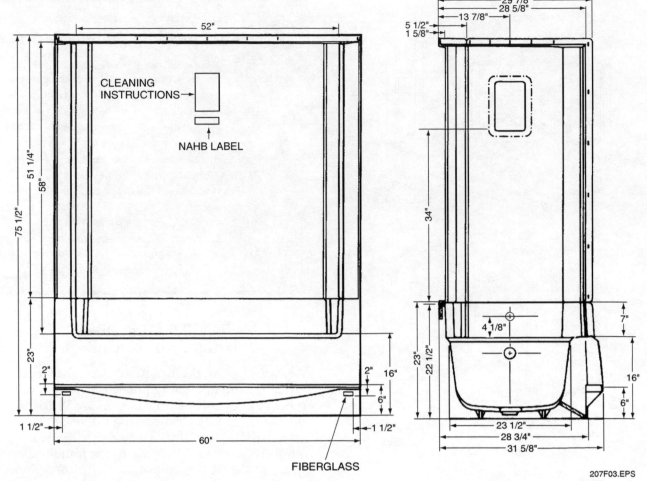

Figure 3 ◆ Rough-in for fiberglass plastic shower and tub unit.

4.1.0 Threaded Valves

Threaded valves have internal threads. To install threaded valves, follow these steps:

Step 1 Apply Teflon tape or **pipe dope** to the coupling threads. If you use pipe dope, do not overfill the threads. Compound buildup in the pipe can cause sanitary problems in the system. Too much compound can accumulate in the valve seat and cause a malfunction.

Step 2 Thread the valve by hand at first. Do this to prevent **cross threading**. Cross threading can damage the soft copper threads on the adapter. After you've started the threads by hand, tighten the valve with a wrench. Hold a second wrench firmly fixed on the adapter to prevent damage to the plumbing already in place.

Step 3 Rotate the valve to a final, vertical position.

Note: When the installation is in-line in a copper system, solder a threaded copper adapter onto the leading end first (see *Figure 4*). Types of threaded valves include temperature/pressure (T/P) relief valves and check valves. Although most check valves are threaded, some may also be soldered.

Using Pipe Dope

A *small* amount of pipe dope on the compression ring will help ensure a watertight fit. Too much and the pipe dope will gum up the openings in the shutoff valve and whatever fixture is eventually attached to it.

Figure 5 ◆ Relief valve.

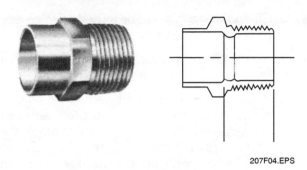

207F04.EPS

Figure 4 ◆ Threaded copper adapter.

4.1.1 Installing Temperature/Pressure Relief Valves

Temperature/pressure (T/P) relief valves are safety devices that prevent a water heater explosion caused by a thermostat failure (see *Figure 5*). The valve opens automatically when the temperature of the water exceeds a preset level. Before you install the valve, remove the protective plastic cap from the relief valve port on the water heater. This port may be on the top or the side of the heater. Apply pipe joint compound or Teflon tape to the valve and thread by hand. Position the valve so that you can easily attach a drain line to the valve.

4.1.2 Installing Check Valves

Check valves provide one-way passage of fluids or gases at specified times. You must be sure to position a check valve properly. Although the proper check valve may have been specified for the job, if you position it improperly, the valve won't function. The check valve in *Figure 6* is designed for installation in both vertical and horizontal lines with upward flow or in any intermediate position. It is important to know the limitations of the valve you're installing. You must also know the direction of flow and make

207F06.EPS

Figure 6 ◆ Check valve.

sure that it corresponds to the directional arrow shown on the check valve. Always match the direction of flow to the arrow. Never rotate the valve on its axis more than 15 degrees. To install check valves, follow these steps:

Step 1 Apply Teflon tape or pipe dope to the male pipe threads. If you use pipe dope, apply it to the first three to four threads. Do not overfill the threads. Compound buildup in the pipe can cause sanitary problems in the system. Too much compound can accumulate in the valve seat and cause a malfunction.

Step 2 Thread the valve by hand at first. Do this to prevent cross threading. Cross threading can damage the threads on the valve. After you've started the threads by hand, tighten the valve with a wrench. Hold a second wrench firmly fixed on the pipe or adapter to prevent damage to the plumbing already in place.

Replace Leaking T/P Valves

Never reuse T/P valves. Always replace any leaking T/P valve. These valves are an essential part of the safety measures built into water heaters and boilers. Many states now require that the T/P valve(s) must be placed on the water heater prior to installation.

4.2.0 Soldered Valves

When soldering valves in copper systems, you must remove the valve interior to prevent damage to the rubber washer by excessive heat buildup. You can replace the interior as soon as the valve assembly has cooled.

4.2.1 Installing a Soldered Shutoff Valve

To install a soldered shutoff valve, follow these steps:

Step 1 Ream the pipe ends and clean the metal with a copper-cleaning tool.

Step 2 Remove the valve interior.

Step 3 Place the **escutcheon** on the **stub-out**. (Do this step if you are installing the valve for a sink, lavatory, or water closet.)

Step 4 Align the valve properly and solder.

Step 5 Replace the valve interior once the valve has cooled.

4.3.0 Solvent-Welded Valves

You'll use solvent welding when you join rigid plastic pipe. You can use a plastic valve or special chrome-plated brass shutoff valve to connect to a **CPVC** (chlorinated polyvinyl chloride) system. The special valves are made with an integral CPVC socket, which is a part of the valve assembly and can accept a solvent weld (see *Figure 7*).

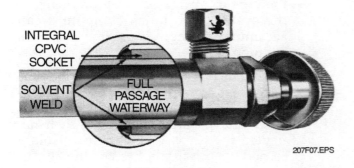

INTEGRAL CPVC SOCKET

SOLVENT WELD

FULL PASSAGE WATERWAY

207F07.EPS

Figure 7 ◆ Integral CPVC socket.

4.3.1 Installing a Solvent-Welded Shutoff Valve

To install a solvent-welded shutoff valve in a bath or kitchen, follow these steps:

Step 1 Cut the cap off the stub end squarely and remove all burrs.

Step 2 Clean the surfaces to be joined with emery paper and a clean cloth.

Step 3 Apply CPVC primer and solvent to both the stub-out and the valve.

Step 4 Push both together while rotating from right to left to ensure complete solvent contact (see *Figure 8*). Carefully align the valve properly now. Once the solvent dries, you won't be able to reposition the valve.

Step 5 Allow adequate drying time as noted on the solvent can.

STUB-OUT   APPLY SOLVENT   CONNECT

207F08.EPS

Figure 8 ◆ Applying solvent to the stub-out and valve ends.

4.4.0 Flanged Valves

Flanged valves (see *Figure 9*) are used where frequent dismantling of piping is necessary. To install flanged valves, follow these steps:

Step 1 Set the valve between the pipe flanges.

Step 2 Put the gasket between the valve flanges and pipe flanges.

Step 3 Line up the bolt holes.

Step 4 Install and tighten the bolts with a **cross pattern**.

Cross Pattern

When you tighten bolts, you apply torque, or pressure, to each bolt. If you were to fully tighten only one bolt at a time, you would run the risk of damage to the fixture because you would apply too much pressure in one spot. In addition, you make it much harder to properly fasten the remaining bolts. To avoid these problems, you must move from bolt to bolt, tightening each bolt a little at a time until all bolts are fully tightened, but not overtightened. You must move between bolts as though you are printing the letter X. This cross pattern of tightening bolts distributes the pressure evenly.

207F09.EPS

Figure 9 ◆ Flanged valve.

4.5.0 Compression Connection Valves

These valves are installed mainly on copper and plastic tubing. The valves consist of three parts—the valve body, the compression ring (also called a **ferrule**), and a compression nut that screws onto the valve body.

CAUTION

Always read and follow the instructions printed on the side of the primer and solvent containers. The manufacturer's safety precautions, handling tips, and drying times are printed there.

4.5.1 Installing Compression Connection Valves

To install a compression connection valve, follow these steps (refer to *Figure 10*):

Step 1 Place the compression nut and then the compression ring on the pipe.

Step 2 Slide the valve onto the pipe until it hits the internal stop.

Step 3 Hold the valve and start threading the compression nut onto the valve.

Step 4 Line up the pipe to the connection to ensure that you do not cross-thread the nut.

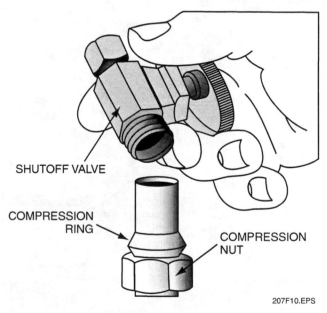

SHUTOFF VALVE

COMPRESSION RING

COMPRESSION NUT

207F10.EPS

Figure 10 ◆ Compression valve.

Cross Threading

Some pipes and fittings are threaded. The threads may be internal or external. When you join an internal thread to an external thread, you must be careful to line up the threads so that they mesh smoothly. If you try to force the items together without ensuring that the threads mesh, you will jam up the connection and could strip the threads.

The compression ring is squeezed between the pipe, the valve, and the nut and compresses onto the pipe. When sufficient pressure is applied, the compression ring cannot be removed from the pipe. This is a very quick and easy joint to make up.

4.5.2 Cartridge Faucets

You can identify a cartridge faucet by the metal or plastic cartridge inside the faucet body. Many single-handle faucets and showers are cartridge designs. Installing a cartridge faucet is fairly easy (see *Figure 11*).

4.5.3 Rotating Ball Faucets

A ball-type faucet has a single handle and is identified by the metal or plastic ball inside the faucet body. The individual parts of a rotating ball faucet are shown in *Figure 12*.

4.5.4 Ceramic Disc Faucets

A ceramic disc faucet is a single-handle faucet that has a wide cylinder inside the faucet body. The cylinder contains a pair of closely fitting ceramic discs that control the flow of water. The top disc slides over the lower disk. The individual parts of the ceramic disc faucet are shown in *Figure 13*.

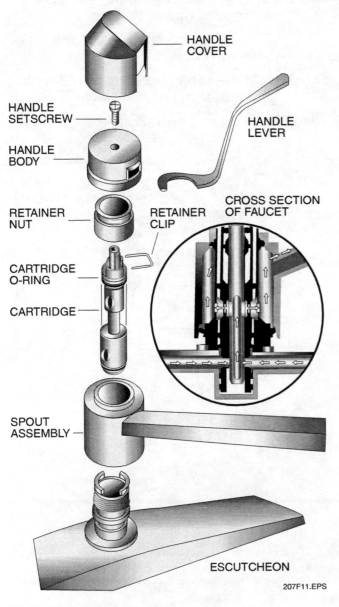

Figure 11 ◆ Cartridge faucet.

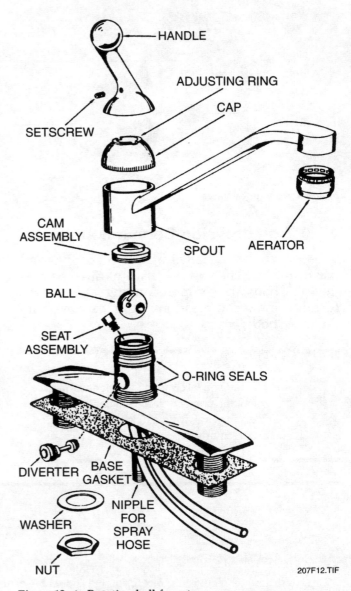

Figure 12 ◆ Rotating ball faucet.

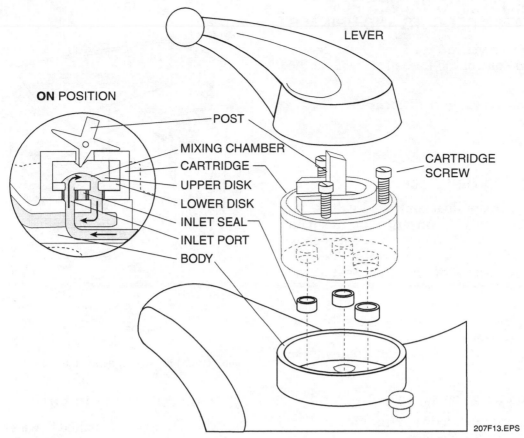

LEVER

ON POSITION

POST

MIXING CHAMBER
CARTRIDGE
UPPER DISK
LOWER DISK
INLET SEAL
INLET PORT
BODY

CARTRIDGE
SCREW

207F13.EPS

Figure 13 ◆ Ceramic disc faucet.

Review Questions

Sections 2.0.0–4.5.4

1. The best way to prevent conflict about who is at fault if fixtures are damaged is to inspect them _____.
   a. just before they are installed
   b. after they are installed on the job
   c. on the day before they are installed
   d. when they are delivered to the job site

2. Bending a fiberglass shower unit during installation _____.
   a. is recommended to test for leaks after installation is done
   b. usually is required to support bearing points properly
   c. may result in cracks that become visible later
   d. will make it more secure in its bracing

3. When installing a threaded valve, a more watertight fit may result if you _____.
   a. wrap several inches of pipe with Teflon tape
   b. put a small amount of pipe dope into the valve seat
   c. apply generous amounts of pipe dope to the threads
   d. use a small amount of pipe dope on the compression ring

4. _____ must be installed in the correct position to operate correctly.
   a. Check valves
   b. Soldered valves
   c. Threaded valves
   d. Shutoff valves

5. Compression connection valves consist of three parts:
   a. the valve body, the compression ring, and the thread adapter
   b. the valve body, the compression ring, and the compression nut
   c. the compression ring, the ferrule, and the washer
   d. the compression ring, the ferrule, and the bolt

5.0.0 ◆ INSTALLING VALVES FOR WATER CLOSETS AND URINALS

Three types of valves are used for water closets and urinals to control the flow of water into the fixture's bowl.

- **Float-controlled valves/ball cocks**
- Flush valves
- **Flushometers**

5.1.0 Installing Ball Cocks

Ball cocks, which are float-controlled valves, are installed in water closet flush tanks and maintain a constant water level in the tank to control flow (see *Figure 14*). When the water level in the tank drops, the valve lifts away from the seat as the float arm lowers. As the water level in the tank increases, the float arm rises and forces the valve down against the seat. This closes the valve and stops the flow of water.

To install the float valve assembly, follow these steps:

Step 1 Turn off the water supply.

Step 2 Lower the threaded base through the bottom hole in the closet tank with the gasket in place against the flange.

Step 3 Place the washer and nut on the base of the assembly on the outside of the tank and tighten. Do not overtighten the nut.

Step 4 Place the riser and coupling nut on the base of the float valve and hand tighten.

Step 5 Align the riser with the shutoff valve. Remove the riser and cut to fit.

Step 6 Reassemble the riser to the tank and valve, and then turn on the water supply.

Step 7 Adjust the float in the tank to achieve the water level indicated on the inside of the tank.

Step 8 Adjust the float with the adjusting screw. Do not bend the float arm to adjust the water level.

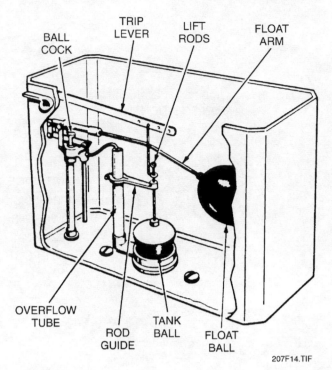

Figure 14 ◆ Float-controlled valve.

5.2.0 Installing Flush Valves

Flush valves also are installed in water closet flush tanks. They control the flow of water from the tank into the bowl (see *Figure 15*). When you push the tank lever, the valve lifts above the tank outlet and floats. This lets the water in the tank flow rapidly into the bowl. When the water level in the tank drops to the point where the valve will no longer float, the valve reseats on the tank drain and the tank refills with water.

To install this valve, follow these steps:

Step 1 Turn off the water supply.

Step 2 Place the flush valve assembly into the center hole in the water closet tank and secure the gasket and lock nut from the outside.

Step 3 Place the spud gasket into the hole on top of the bowl and lower the tank into place.

Step 4 Place tank bolts with gaskets through holes in the tank's interior and bowl and install washers and nuts. Tighten firmly, but do not overtighten. Overtightening can damage the tank.

 CAUTION

Never overtighten the nut on the base of the assembly on the outside of the tank. Overtightening puts stress on the fixture and can cause it to crack.

 CAUTION

Never bend the float arm to adjust the water level. The rod may work around a half-turn (180 degrees) and cause the tank to overflow.

5.3.0 Installing Flushometers

Plumbers install flushometers for water closets or urinals. They don't require a storage tank because the water flows, under pressure, directly into the fixture. The scouring action of the water created by the flushometer generally cleans more effectively than the gravity flow from a storage tank.

The flushometer is popular in commercial installations for two reasons:

- It can be connected directly to the water supply pipe.
- It can be flushed repeatedly without waiting for a storage tank to refill.

Flushometers are available in either a diaphragm type (see *Figure 16*) or a piston type. The diaphragm type stops the flow of water when water pressure in the upper chamber forces the diaphragm against the valve seat. Motion of the handle in any direction pushes the plunger against the auxiliary valve. Even the slightest tilt of the handle will transfer water from the upper chamber around the diaphragm.

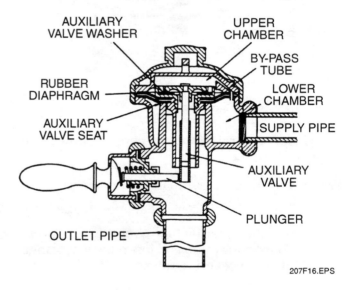

Figure 16 ◆ Diaphragm type flushometer.

When pressure decreases in the upper chamber the diaphragm rises, allowing water to flow into the fixture. While the fixture is in the flushing mode, a small amount of water flows through the by-pass. As the water fills the upper chamber, the diaphragm is forced to reseat against the valve seat. This action shuts off the flow of water to the fixture. The piston-type flushometer works in much the same way the diaphragm type works.

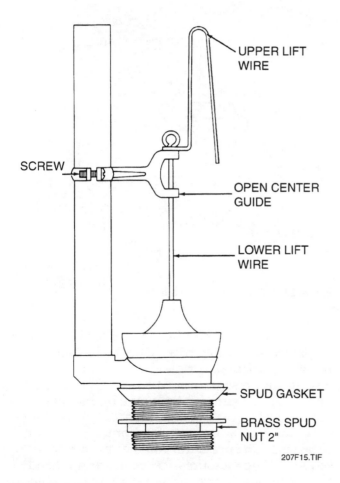

Figure 15 ◆ Flush valve.

You should install the flushometer with a control stop and a vacuum breaker. Install the control stop (see *Figure 17*) in the water supply line serving the flushometer. The control stop regulates the flow of water entering the flushometer. It also allows you to shut off the water supply to individual fixtures to make repairs. Install the vacuum breaker (see *Figure 18*) between the flushometer outlet and the flushometer tube. The vacuum breaker prevents back siphoning of polluted water into the water supply piping.

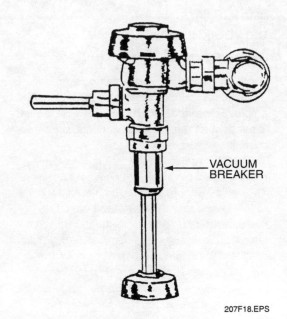

Figure 18 ◆ Vacuum breaker.

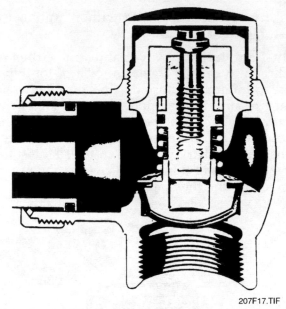

Figure 17 ◆ Control stop.

To install control stop connections, follow these steps:

Step 1 Cut the stub-out to its proper length.

Step 2 Solder the male sweat adapter to the tube.

Step 3 Slip the covering tube over the male sweat adapter and stub and insert the wall flange over the covering tube.

Step 4 Lock the setscrew over the covering tube to prevent it from sliding back into the wall cavity.

Step 5 Thread the angle stop to the stub-out.

Step 6 Attach the flushometer (see *Figure 19*). Do this carefully to avoid damaging the chrome exterior on the coupler nut and on other parts of the valve.

Step 7 Install the vacuum breaker, which is usually a part of the fixture connector.

Step 8 Secure the connector to the valve and fixture.

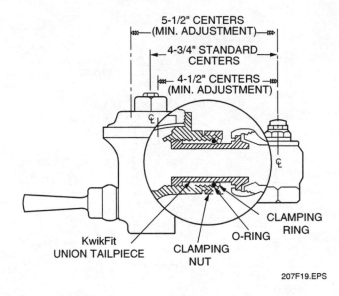

Figure 19 ◆ Flushometer installation.

You can adjust the flushometer for both water consumption and noise. Check the manufacturer's instructions for more information. You can also make pressure adjustments. Find the ideal operating pressure for the valve by flushing and adjusting. As you gain experience you will know how much to adjust the flushometer to get the ideal operating pressure. You may need a pressure reducer if line pressure exceeds the recommended pressure. Low pressure is also a problem. Although many valves can function at a low pressure (minimum 15 pounds per square inch), they may not function well.

6.0.0 ◆ INSTALLING LAVATORIES, SINKS, AND POP-UP DRAINS

Installing lavatories is very similar to installing sinks. Either of two configurations is possible in the installation process. The fixtures may be either wall-hung or **built-in**, which affects how you support them. Built-in fixtures are placed in a countertop, and you may have to cut the opening.

6.1.0 Wall-Hung Lavatories and Sinks

The following installation steps are provided as a general guide. *Always* read and follow the manufacturer's instructions to determine the specific instructions for each fixture.

The steps for installing wall-hung lavatories and sinks are as follows:

Step 1 Check the fixture for damage.

Step 2 Check the rough-in dimensions (see *Figure 20*).

Step 3 Confirm that the blocking is located in the correct position to receive the bracket.

Step 4 Secure the bracket to the framing.

Step 5 Place the lavatory or sink on the bracket temporarily and check it for level.

Step 6 Remove the fixture and adjust the bracket as necessary.

Step 7 Install the faucets, drain, and any other trim. Additional information about this part of the task is included in the manufacturer's directions supplied with the parts.

Step 8 Secure the fixture to the bracket and seal the joint between the wall and the fixture with the appropriate sealer.

Step 9 Attach the water supply.

Step 10 Install the waste piping.

Step 11 Turn on the water and test for leaks.

Step 12 Turn off the water and cover the fixture to protect it from damage.

6.2.0 Built-In Lavatories and Sinks

The following installation steps are provided as a general guide. *Always* read and follow the manufacturer's instructions to determine the specific instructions for each fixture.

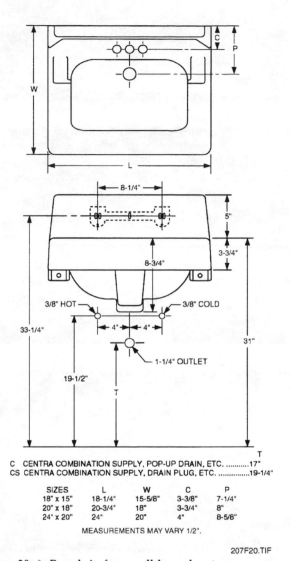

| C | CENTRA COMBINATION SUPPLY, POP-UP DRAIN, ETC. | ....17" |
| CS | CENTRA COMBINATION SUPPLY, DRAIN PLUG, ETC. | ....19-1/4" |

SIZES	L	W	C	P
18" x 15"	18-1/4"	15-5/8"	3-3/8"	7-1/4"
20" x 18"	20-3/4"	18"	3-3/4"	8"
24" x 20"	24"	20"	4"	8-5/8"

MEASUREMENTS MAY VARY 1/2".

207F20.TIF

Figure 20 ◆ Rough-in for a wall-hung lavatory.

DID YOU KNOW?

A Strong and Reliable Material

Porcelain enamel is used for most sinks and lavatories because it is sanitary. Its nonporous surface resists bacteria growth and cleans easily. Because it is a nontoxic mineral substance, it is environmentally friendly. In addition, porcelain is—

- Abrasion resistant
- Chemical resistant
- Corrosion resistant
- Weather resistant
- Graffiti-proof
- Flame-proof
- Fade-proof

207F21.EPS

Figure 21 ◆ Rim-mounted, under-the-counter, and self-rimming lavatories.

CAUTION

Protect the countertop from scratches. Make certain the base of your saw is free of burrs and sharp edges.

Built-in lavatories and sinks are placed in an opening cut into a cabinet's countertop. Three styles are available (see *Figure 21*):

- **Rim-mounted**
- **Under-the-counter**
- **Self-rimming**

Although a cabinetmaker usually cuts the opening in the countertop, you may have to do this job. Before you cut out the opening, make sure that the cabinet's structural parts won't interfere with the installation. You'll need enough space left so that you can tighten the clips that fasten the fixture to the bottom of the countertop.

The steps for installing built-in lavatories and sinks are as follows:

Step 1 Measure the correct position and trace the opening on the countertop. Most manufacturers provide a pattern for this step. If you don't have a pattern, turn the fixture over, position it on the countertop, and carefully trace a line around the metal rim. If you use this process, you must make allowances for the rim, otherwise the cutout will be too big.

Step 2 Place the fixture to one side.

Step 3 Drill a starter hole at a point along line where you would logically begin cutting. Insert the blade of your saber saw into the hole and make the cutout.

Step 4 Test fit the rim in the opening. If the fixture is self-rimming, insert the fixture in the opening to test the fit.

Step 5 Make any adjustments necessary to the shape of the opening.

Step 6 Install the faucets and other trim.

Step 7 Apply fixture sealant uniformly to the rim. Do this step carefully and completely to ensure that water will not flow under the fixture.

Step 8 Place the fixture in the opening and adjust the position.

ON THE LEVEL

Installation Tip

Install the faucets before you install the sink or lavatory. This makes it easier for you to reach the nuts on the underside. Once the fixture is installed, there won't be much room between the bottom of the bowl and the rear of the cabinet to use tools.

Step 9 Install and tighten the clamps that hold the fixture in place from underneath the countertop. Tighten the clips uniformly.

Step 10 Connect the water supply and the waste piping.

Step 11 Turn on the water and check for leaks.

6.3.0 Installing Pop-Up Drains

Most lavatories come with **pop-up drains.** These are more attractive than rubber stoppers and won't get lost. They come in finishes to match the faucets. To install this drain, follow these steps:

Step 1 Attach the pivot rod assembly to the lift rod (see *Figure 22*).

Step 2 Adjust the length of the **lift rod assembly** by loosening the clevis screw and moving the clevis so that the pop-up plug is closed when the lift knob is up and open when the lift knob is down.

Step 3 Check the retaining nut that secures the pivot rod for leaks.

Step 4 Tighten the retaining nut slightly to correct a leak, but don't overtighten. If the nut is too tight, the pop-up plug may be difficult to operate.

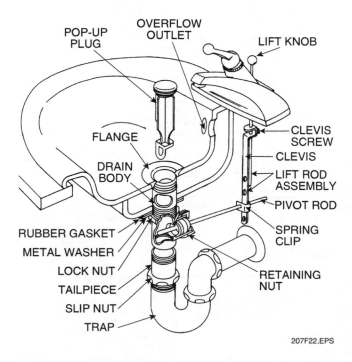

207F22.EPS

Figure 22 ◆ Typical pop-up drain assembly.

DID YOU KNOW?
American Inventors and the Water Closet

The first Americans awarded a patent for a water closet were James Henry and William Campbell. In 1875, they developed a plunger closet. This unit was not very sanitary, and some of the industry's earliest pioneers shunned it.

Patent grants for water closets increased as inventors realized the potential market for an improved model. John Mann got a patent for his three-pipe siphonic closet in 1870. In 1876, William Smith earned his for a jet siphon closet. Thomas Kennedy improved on Mann's design and patented a siphonic closet that needed only two delivery pipes. One flushed the rim and the other started the siphon. More improvement occurred in 1890 with William Howell's water closet. His model eliminated the lower trap.

The U.S. Patent Office received applications for 350 new water closet designs between 1900 and 1932. Two of the first were granted to Charles Neff and Robert Frame. These New Englanders produced a siphonic wash-down closet. But their unit's bowl sometimes overflowed. Around 1910, Fred Adee redesigned that bowl, eliminating the messy overflows, and gave birth to the siphonic closet in America.

In the early 1900s, the U.S. Patent Office granted patents for the flushometer valve, a backflow preventer, a wall-mounted closet with a blowout arrangement, a tank that rests on the bowl, and reverse trap toilets.

7.0.0 ◆ INSTALLING WATER CLOSETS

Water closets come in a variety of styles and materials. The following installation instructions are a general guide. Remember to always read and follow the manufacturer's instructions.

Before installing the water closet, check the condition of the floor. Don't set the water closet in place until a nonporous floor covering is installed. In older buildings, you may have to replace part of the floor. The floor must carry the weight of the fixture and the person using it, so be sure that it is solidly constructed.

CAUTION

Never overtighten the bolts on the bowl or tank. Overtightening puts stress on the fixture and can cause it to crack.

To install a water closet, follow these steps:

Step 1 Check the rough-in opening and closet flange dimensions.

Step 2 Check the bowl or tank for defects and cracks.

Step 3 Set the **closet bolts** in the closet flange with the threads up.

Step 4 Turn the bowl bottom up and position the **wax seal** (see *Figure 23*).

Step 5 Apply a bead of fixture sealant along the outside rim of the bowl.

Step 6 Set the bowl over the closet flange. Make certain the closet bolts extend through the holes in the base of the bowl.

Step 7 Press the bowl firmly into position. Apply your full weight to ensure that the wax ring is completely sealed (see *Figure 24*).

Step 8 Check the bowl for level. Shim as required to level the bowl. If you use shims, grout the space between the bottom of the bowl and the floor to support the bowl evenly.

Step 9 Secure the bowl with the closet bolts. Stop tightening the bolts when you can no longer rock the fixture. Don't overtighten the bolts.

Step 10 Cut off the excess bolt length and install the bolt caps.

Step 11 Check the tank to ensure that the ball cock and flush valve are completely and properly installed.

Step 12 Install the tank and hand tighten the tank bolts. Follow the manufacturer's directions to make sure that you install the washers and seals correctly.

Step 13 Make sure the tank is level.

Step 14 Tighten the tank bolts until the tank is secure. Do not overtighten.

Step 15 Hook up the water supply and fill the tank.

Step 16 Flush the water closet several times. Adjust the float valve to control the water level and check the entire assembly for leaks.

Step 17 Install the closet seat securely.

Step 18 Turn off the water, cover the fixture, and post a DO NOT USE sign to protect the fixture until the project is finished.

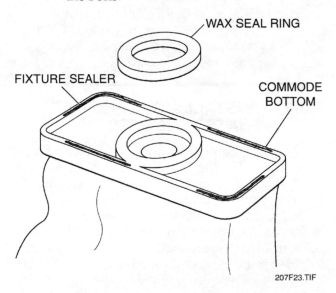

Figure 23 ◆ Location of the wax seal.

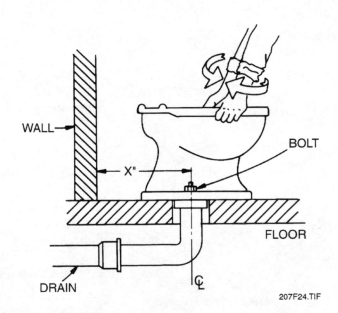

Figure 24 ◆ Seating the bowl.

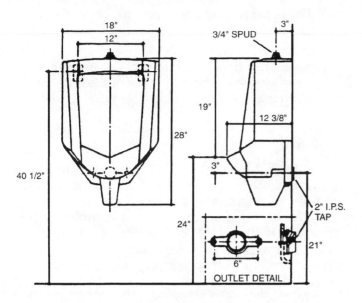

207F25.TIF

Figure 25 ◆ Rough-in of a wall-hung urinal.

8.0.0 ◆ INSTALLING URINALS

As with the other fixtures covered in this module, you must read and follow the manufacturer's instructions to perform a proper installation and protect the warranty. The installation steps included here are a general guide. You'll need a flushometer and a drain to complete the installation of a wall-hung urinal (see *Figure 25*).

To install a wall-hung urinal, follow these steps:

Step 1 Unpack and inspect the urinal for damage, defects, and cracks.

Step 2 Install the drain fitting.

Step 3 Install hangers at the correct height, if they are required. Otherwise, locate bolts to fasten the fixture to the wall framing.

Step 4 Bolt the urinal to the wall, making certain that it is level.

Step 5 Install the flush valve. Work carefully to avoid scratching or otherwise damaging the chrome-plated parts.

Step 6 Connect the waste piping.

Step 7 Turn the water on and adjust the flushometer.

Step 8 Flush the unit several times to confirm that it works correctly.

Step 9 Turn off the water, cover the fixture, and post a DO NOT USE sign to protect the fixture until the project is finished.

 DID YOU KNOW?

Thomas Crapper

Some people say there was no such person as Thomas Crapper and that he never invented the flush toilet. But he did exist. He had a successful career in the plumbing industry in England from 1861 to 1904. When he retired in 1904, he sold his shop to two partners who operated the company under the Crapper name until its closing in 1966.

However, he did not invent the "Silent Valveless Water Waste Preventer." This was a siphonic discharge system that allowed a toilet to flush effectively when the cistern was only half full. A British patent for this invention was issued to a man named Albert Giblin. There are a couple of theories on how Thomas Crapper came to be associated with this device. One is that Giblin worked for Crapper and authorized his use of the product. The more likely theory, though, is that Crapper bought the patent rights from Giblin and marketed the product himself. Whichever theory is true, Thomas Crapper's name has become linked with the invention of the flush toilet.

Review Questions

Sections 5.0.0–8.0.0

1. Overtightening the nut on the base assembly of a water closet on the outside of the tank _____.
   a. is impossible
   b. is required for most water closets
   c. may cause the fixture to crack from stress
   d. may prevent leaks from the shutoff valve

2. A _____ delivers water to a urinal or water closet without a storage tank.
   a. flushometer
   b. flush valve
   c. float valve
   d. pressure valve

3. A built-in sink or lavatory should be installed _____.
   a. at the same time as the shower or bathtub
   b. as soon as the cabinet is available
   c. before faucets are installed
   d. after faucets are installed

4. The _____ regulates the flow of water entering the flushometer.
   a. piston
   b. diaphragm
   c. control stop
   d. vacuum breaker

5. The pivot rod retaining nut on a pop-up drain should be installed _____.
   a. as tightly as possible
   b. just tightly enough to prevent leaking
   c. as loosely as possible without falling off
   d. so that the plug opens when the lift knob is down

Summary

Installing fixtures, valves, and faucets is your finish work. Once the walls are closed up, it's the one part of your plumbing job that everyone sees. Doing this work neatly, accurately, and completely the first time will prevent costly callbacks. Your work will be viewed as competent and professional, and this will reflect well on you and your company.

To do a professional job, protect all fixtures before, during, and after installation. Check the job specifications to ensure that the right fixtures, with all parts intact, arrive at the job site. Always read and follow the manufacturer's instructions to ensure a proper installation and to protect the warranty. Inspect and test installed fixtures to ensure that they are intact, work properly, and don't leak. Finally, clean up your work area. Remove and dispose of unused materials and other waste. Your co-workers, the trades coming in after you, and especially your customers will appreciate this professional courtesy.

Trade Terms Introduced in This Module

Ball cock: A type of float-controlled valve that controls the flow of water in a water closet.

Blocking: Pieces of wood used to secure, join, or reinforce structural members or to fill spaces between them.

Built-in: A type of fixture designed for installation in a cabinet or frame such as a lavatory or sink.

Callback: A term that refers to being called back to the worksite to repair or replace defective materials or to correct faulty workmanship.

Closet bolt: A bolt with a large diameter, low circular head that is cupped on the underside so that it is sealed against the surface when the bolt is tightened. It is used to fasten a water closet bowl to the floor.

CPVC: An abbreviation for chlorinated polyvinyl chloride, which is a plastic used for piping in hot- and cold-water systems and in drainage systems.

Cross pattern: A method for tightening bolts or screws in an even sequence, tightening each bolt or screw a little at a time to avoid putting too much pressure on any one bolt or screw.

Cross threading: A condition that occurs when the initial female thread on a pipe or fitting does not properly mesh with the initial male thread. This causes the connection to jam and can strip the threads.

Escutcheon: A flange on a pipe used to cover a hole in a floor or wall through which the pipe passes.

Ferrule: A metal sleeve fitted with a screwed plug.

Finish work: The completion phase of any construction project. Generally the most visible part of the work, it includes paint, stain, trim, and fixtures.

Float-controlled valve: A valve that controls water flow. When the water level drops, the valve lifts and allows water to flow. When the water level rises, the valve closes and stops the flow of water.

Floor-mounted: A type of fixture designed for installation on the floor, such as a bathtub or a water closet.

Flushometer: A valve that allows a preset amount of water to enter a fixture, such as a water closet or a urinal, to flush it.

Lift rod assembly: A mechanism consisting of the clevis and clevis screw that is used to adjust the operation of a pop-up valve.

Pipe dope: A compound used to lubricate a pipe joint and to make it leakproof.

Pop-up drains: A mechanism that allows a user to open or close a lavatory drain by pulling on a lift knob connected to a pivot rod.

Rim-mounted: A type of sink or lavatory that is framed with a metal rim.

Rough-in: The initial stage of any phase of construction. The earliest stage of plumbing installation.

Self-rimming: A type of sink or lavatory that fits directly on the countertop.

Stub-out: The part of the pipe that sticks through the wall or floor and to which a valve, trap, or fixture is attached.

Tub waste: The drain fixture and fittings that take wastewater from the tub.

Under-the-counter: A type of sink or lavatory that is installed under the opening in the countertop for a smooth, easy-to-clean surface.

Wall-hung: A type of fixture designed to be hung from bolts or carriers set in the wall—for example, lavatories and urinals.

Wax seal: A ring made of heavy-duty wax or rubber that fits between the bottom of the water closet bowl and the floor.

Answers to Review Questions

Sections 2.0.0–4.5.4
1. d
2. c
3. d
4. a
5. b

Sections 5.0.0–8.0.0
1. c
2. a
3. d
4. c
5. b

Figure Credits

Brass Craft	207F07, 207F08
Crane Company	207F09
Fluidmaster	207F14
Mansfield Plumbing Products	207F05
Mueller Brass Company	207F04
NIBCO	207F06
Owens/Corning Fiberglas	207F03, 207F20
Plumbers and Pipefitters Library	207F15
Sloan Value Company	207F17, 207F19

NCCER CRAFT TRAINING USER UPDATES

The NCCER makes every effort to keep these textbooks up-to-date and free of technical errors. We appreciate your help in this process. If you have an idea for improving this textbook, or if you find an error, a typographical mistake, or an inaccuracy in the NCCER's Craft Training textbooks, please write us, using this form or a photocopy. Be sure to include the exact module number, page number, a detailed description, and the correction, if applicable. Your input will be brought to the attention of the Technical Review Committee. Thank you for your assistance.

Instructors – If you found that additional materials were necessary in order to teach this module effectively, please let us know so that we may include them in the Equipment and Materials list in the Instructor's Guide.

Write: Curriculum Revision and Development Department
                 National Center for Construction Education and Research
                 P.O. Box 141104, Gainesville, FL 32614-1104

Fax: 352-334-0932

E-mail: curriculum@nccer.org

Craft _____ Module Name _____

Copyright Date _____ Module Number _____ Page Number(s) _____

Description _____

_____

_____

_____

(Optional) Correction _____

_____

_____

(Optional) Your Name and Address _____

_____

_____

Installing
Water Heaters

COURSE MAP

This course map shows all of the modules in the second level of the Plumbing curriculum. The suggested training order begins at the bottom and proceeds up. Skill levels increase as you advance on the course map. The local Training Program Sponsor may adjust the training order.

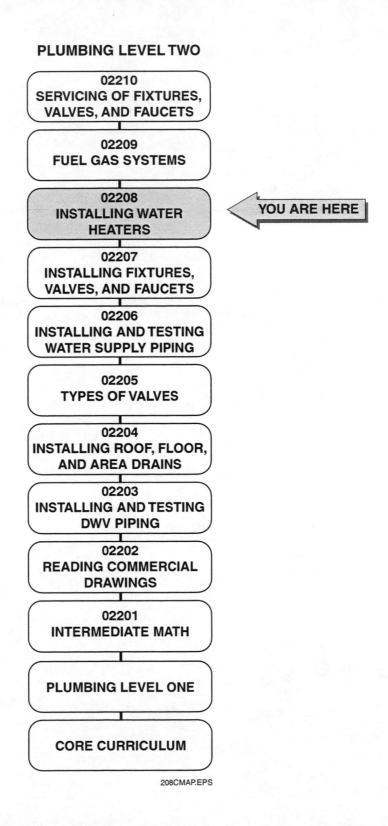

PLUMBING LEVEL TWO

02210
SERVICING OF FIXTURES, VALVES, AND FAUCETS

02209
FUEL GAS SYSTEMS

02208
INSTALLING WATER HEATERS

YOU ARE HERE

02207
INSTALLING FIXTURES, VALVES, AND FAUCETS

02206
INSTALLING AND TESTING WATER SUPPLY PIPING

02205
TYPES OF VALVES

02204
INSTALLING ROOF, FLOOR, AND AREA DRAINS

02203
INSTALLING AND TESTING DWV PIPING

02202
READING COMMERCIAL DRAWINGS

02201
INTERMEDIATE MATH

PLUMBING LEVEL ONE

CORE CURRICULUM

208CMAP.EPS

MODULE 02208 CONTENTS

Figures

Table

Installing Water Heaters

Objectives

When you have completed this module, you will be able to do the following:

1. Describe the basic operation of water heaters.
2. Identify and explain the functions of the basic components of water heaters.
3. Install an electric water heater.
4. Install a gas water heater.
5. Describe the safety hazards associated with water heaters.

Prerequisites

Before you begin this module, it is recommended that you successfully complete the following modules: Core Curriculum; Plumbing Level One; Plumbing Level Two, Modules 02201 through 02207.

Required Trainee Materials

1. Paper and sharpened pencil
2. Appropriate Personal Protective Equipment

1.0.0 ◆ INTRODUCTION

Most residential and light commercial structures have centrally located water heating units. These units heat, store, and supply hot water. Fixtures that use hot water include showers, lavatories, sinks, bathtubs, dishwashers, and washing machines. A piping system brings the cold water supply to the water heater and delivers heated water to the hot water fixtures. Gas, oil, electricity, or solar power may provide the energy to heat the water. Gas heaters require a piping system that delivers fuel to the burner unit and a venting system that carries away smoke and other combustion products.

As a plumber, you will install the piping for the water heater, place the unit in operation, and perform minor maintenance. This module will give you a basic understanding of how water heaters operate and their special plumbing needs.

2.0.0 ◆ BASIC OPERATION OF WATER HEATERS

Most modern water heating units combine a heating unit with a storage tank (see *Figure 1*). These units automatically provide a ready supply of hot water at a preset temperature. To operate properly, a water heater needs a heat source, temperature controls, and safety devices (see *Figure 2*).

208F01.EPS

Figure 1 ◆ Water heater.

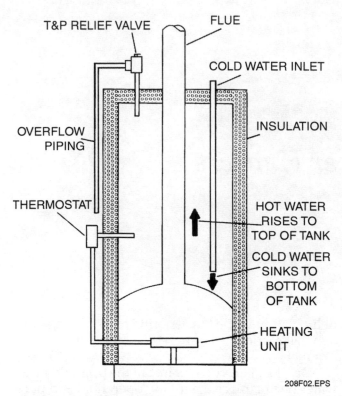

Figure 2 ◆ Basic operation of a gas water heater.

Cold water enters the tank near the bottom, and hot water is drawn off near the top. The insulated water tank design takes advantage of water's natural properties. As water heats it expands, becomes lighter, and rises to the top of the tank. Denser (heavier) cold water sinks to the bottom of the tank where it is heated, and the cycle repeats itself. This circulation of water in the storage tank provides a constant reservoir of hot water. The size of the storage tank selected for an installation depends on the peak demand for hot water in the building.

3.0.0 ◆ TYPES OF WATER HEATERS

Water heaters are of two basic types. One type is the gas-fired water heater, which is fueled by natural gas or propane (see *Figure 3*). The other type is the electric water heater, which is fueled by electricity (see *Figure 4*). The most commonly installed types, whether gas or electric, are automatic storage tank water heaters. These heat and store water at a thermostatically controlled temperature for delivery on demand.

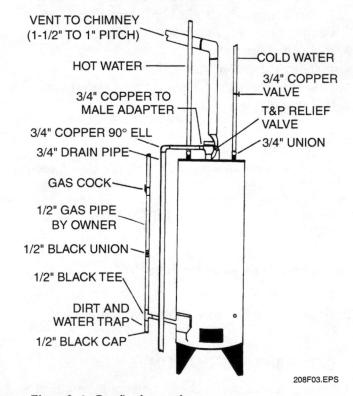

Figure 3 ◆ Gas-fired water heater.

ON THE · LEVEL ·

Relief Valve Drip

All water heaters have a degree of expansion. Many relief valves drip to equalize the pressure. In most cases, installing a small expansion tank solves this drip problem.

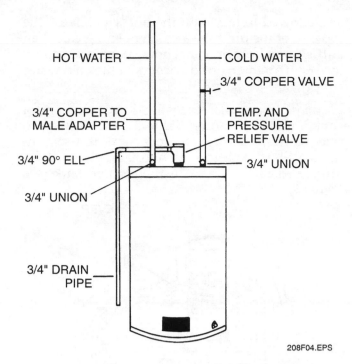

HOT WATER — — COLD WATER

3/4" COPPER VALVE

3/4" COPPER TO
MALE ADAPTER

TEMP. AND
PRESSURE
RELIEF VALVE

3/4" 90° ELL

3/4" UNION

3/4" UNION

3/4" DRAIN
PIPE

208F04.EPS

Figure 4 ◆ Electric water heater.

A third type of water heater uses solar energy, and several types of solar heaters are available. The most practical ones include a backup system that uses conventional fuels for overcast days or when hot water demand is higher than normal.

3.1.0 Gas Automatic Storage Tank Water Heaters

Natural gas and liquefied petroleum gas (LPG) are the fuels most commonly used for gas-fired storage tank water heaters. (Oil is a third fuel type, but oil-fired hot water heaters are not common.) The fuel burns in a heating unit that contains a firebox located under the tank inside the outer metal jacket. This burning, or combustion, produces byproducts that travel up a **flue** to a chimney or outdoors. The flue is a tube-like passage that runs up the center of the heating tank. **Baffles** inside the flue collect heat from the gases moving through the flue and transfer it to the water in the tank.

Heaters using LPG have smaller fuel and heating orifices (openings) than natural gas heaters. This design aids in handling a more concentrated fuel. Although liquefied petroleum is stored in large tanks as a liquid, it vaporizes before it is burned.

 DID YOU KNOW?

Solar Thermal Systems

Solar thermal systems use the sun's energy to heat water or produce power. Small-scale water heating systems use flat plate collectors to capture the sun's energy. Power plants must concentrate the sunlight for the high temperatures they need to produce power.

A home hot water system can be retrofitted with a solar thermal system. Most of these systems have a backup heating element to fill in on cloudy days or to cover a higher than normal hot water use. A solar thermal system can take care of 60 percent to 80 percent of residential hot water needs.

About 250,000 commercial and industrial buildings in the United States use these systems to provide hot water or space heating. Industries in which these systems are most common include laundry, food service, food processing, metal plating, and textiles.

3.1.1 Gas Water Heater Components

Following are the components commonly found in gas-fired water heaters:

Anode – Often called a *sacrificial anode*, this is a metal rod that helps to keep the tank from corroding. Electrolysis eats away the anode instead of the tank. The anode is screwed into the top of the water heater. You must replace worn anodes to preserve the tank. Sometimes the anode is built into a special hot water outlet fitting. Softened water can cause the anode to wear out more quickly.

Jacket and *top cover* – Made of steel finished in baked enamel, they form the outer shell of the water heater.

Insulation – Located between the tank and the outer shell, it reduces heat loss from the water in the storage tank.

Storage tank – Commonly referred to as a *center-flue tank*, it allows heated flue gases to cover the bottom of the tank before they enter the vertical flue. A baffle inside the flue slows the flow of hot combustion gases and maximizes heat transfer to the water stored in the tank.

The inside surface of the storage tank is lined with a glass coating fused to the surface. This coating protects the tank from the corrosive effects of minerals in the hot water. Minerals that build up on the walls create **scale**, which acts as an insulator. This, in turn, reduces heat transfer, resulting in a longer run time.

The top of the storage tank includes three openings. The cold inlet opening connects to the cold water supply pipe. The hot water outlet opening connects to the hot water supply pipe. Relief valves are installed on all tank-type water heaters. The common type gives relief under pressure and temperature.

Temperature/pressure (T/P) relief valve – A safety device, it is installed in the storage tank or as close to the tank as possible (see *Figure 5*). It prevents the tank from exploding in case the thermostat fails to operate properly and the water becomes overheated. A sensor in the valve detects an extreme rise in temperature. When the temperature increases, pressure builds up in the tank and opens the valve. This releases water or steam and equalizes the excess pressure in the tank. To test the valve, lift the lever. If water is not released, replace the valve. A **relief valve opening** provides access for installation of a T/P relief valve.

Water heater control and *tank drain valve* – These are attached to openings on the side of the storage tank. The water heater control adjusts the water temperature. The tank drain valve is located as close to the bottom of the tank as possible. Customers use this valve to drain the tank and clean out sediment and scale. Many manufacturers recommend this preventive maintenance. Consult the manufacturer's manual for the recommended schedule.

Dip tube – A device that prevents cold water from mixing with hot water, it delivers incoming cold water through the stored hot water to the bottom of the tank. The cold water is then rapidly heated

and allowed to mix with the hot water. Near the top end of the dip tube a small opening called an **anti-siphon hole** prevents hot water from siphoning out if an interruption occurs in the cold water supply.

Thermostatic control – A sensing device, the control contains a temperature dial that permits users to select a desired water temperature. It also contains a thermostatic sensing element, or *probe*, immersed in the water in the storage tank (see *Figure 6*).

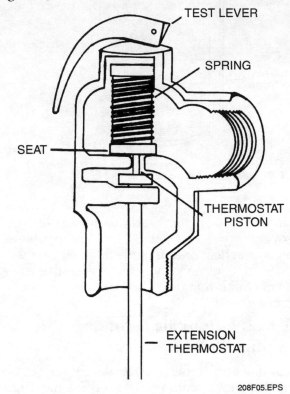

208F05.EPS

Figure 5 ◆ Temperature/pressure relief valve.

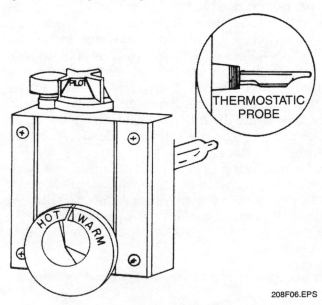

208F06.EPS

Figure 6 ◆ Thermostatic probe.

When a customer draws hot water from the tank, cold water flows in. The sensor in the control activates the gas (see *Figure 7*). It does this by opening a valve and allowing gas to flow to the burner. Once the cold water heats to a preset temperature, the thermostat closes the valve and shuts off the gas supply.

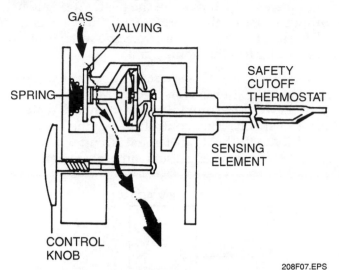

208F07.EPS

Figure 7 ◆ Operation of the thermostatic probe.

Main burner – This is the chamber in which the fuel gas mixes with air and burns. The combustion in the chamber heats the water in the tank to a preset temperature.

Safety pilot – Also called a **pilot burner**, it ignites the gas at the main burner when turned on by the thermostat (see *Figure 8*). It is called a safety pilot because the gas supply to the water heater automatically shuts off if the pilot goes out.

Thermocouple – A small electric generator made of two different metals joined firmly together, it produces a small electric current generated by the heat from the safety pilot. This current holds the safety shutoff gas valve in an open position. If the pilot flame burns too low or goes out, the current breaks and the valve spring releases, shutting off the gas flow to the main burner.

Valve control unit – This unit combines the automatic pilot valve, or **solenoid valve**, and the thermostat valve. The safety pilot light controls the solenoid valve. The thermostat controls the thermostat valve.

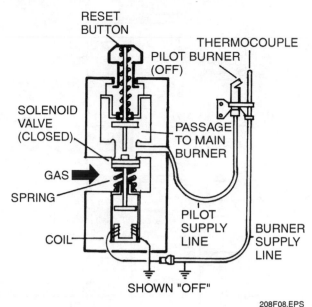

208F08.EPS

Figure 8 ◆ Safety pilot.

3.2.0 Electric Automatic Storage Tank Water Heaters

In an electric automatic storage tank water heater, an element immersed in water heats the water in the storage tank. Most electric water heaters require a 240-volt electrical service to operate the heating element efficiently. Although some water heaters will operate on a 120-volt electrical service, they usually are not as efficient.

3.2.1 Electric Water Heater Components

Except for the flue, baffles, main burner, safety pilot, thermocouple, and valve control unit, the components commonly found in an electric water heater are identical to those found in gas water heaters. Additional components in electric water heaters include the following:

Automatic thermostat – A device that controls water temperature, the automatic thermostat senses the tank's outside surface temperature and starts or stops the flow of electricity to the heating elements (see *Figure 9*). Other types of electric water heaters use a thermostat exposed to the water with a remote sensing bulb located in an immersion well.

Keep Relief Valves Clear

A relief valve on a water heater in a school building was plugged off. The resulting explosion blew out a brick wall. Always check to ensure that relief valves are operating correctly and that they are clear of obstructions.

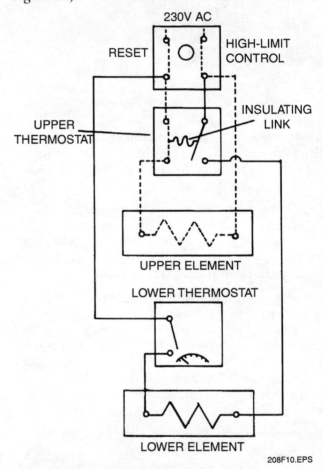

Immersion elements – See *Figure 10*. The diagram shown includes two heating elements. In normal operation, only the lower element operates. When hot water demand is high, only the top element—which is larger and produces more heat—operates. An automatic thermostat controls both immersion elements. Water carries away the heat generated by the elements. If you **dry fire** the heater (operate it without water in the storage tank) the elements will burn out. *Never* turn on an electric water heater unless water fills the storage tank.

High-limit control – This is a safety device that protects against extreme water temperatures caused by defective thermostats or grounded water heater elements. This control breaks the flow of electricity to the heating element circuit when the tank surface reaches a preset temperature (refer to *Figure 10*).

230V AC

RESET HIGH-LIMIT CONTROL

INSULATING LINK

UPPER THERMOSTAT

UPPER ELEMENT

LOWER THERMOSTAT

LOWER ELEMENT

208F10.EPS

Figure 10 ◆ Wiring diagram.

Figure 9 ◆ Automatic thermostat on an electric water heater.

208F09.EPS

ON THE ◆ LEVEL ◆

Tankless Water Heaters

Tankless water heaters have a flow switch. If the switch doesn't detect maximum flow, the heater will not heat. Be sure to clean the aerator on the sink or install a larger gallons-per-minute aerator to solve this problem.

4.0.0 ◆ INDIRECT WATER HEATERS

In an **indirect water heater**—also called a *heat generator* or *heat exchanger*—service water enters the boiler. The boiler water passes through a copper coil and the heated water is delivered to the fixtures (see *Figure 11*). Latent heat from the hot water is then captured by the heat exchanger and mixes with cold water coming into the boiler. This is an efficient way to heat water. Indirect heaters are usually found in commercial applications. The high-limit aquastat is a safety device that turns the boiler off if the boiler water reaches the maximum temperature.

Figure 12 shows a boiler with and without a tankless heater. The components of this system include a heat exchanger and a circulator/adapter. In this system, the boiler heats the water passing through the heat exchanger and the water heater serves only as a storage tank. When the system includes a tankless heater, the heat exchanger is not required because the tankless heater functions as the heat exchanger.

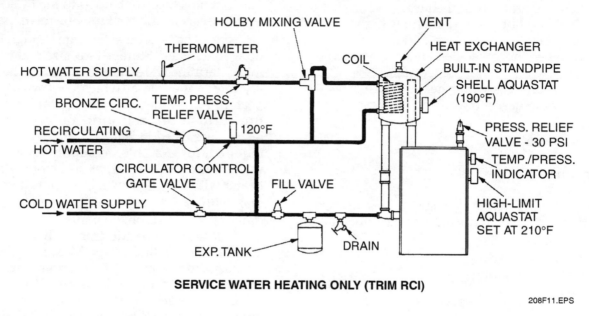

Figure 11 ◆ Indirect water heater.

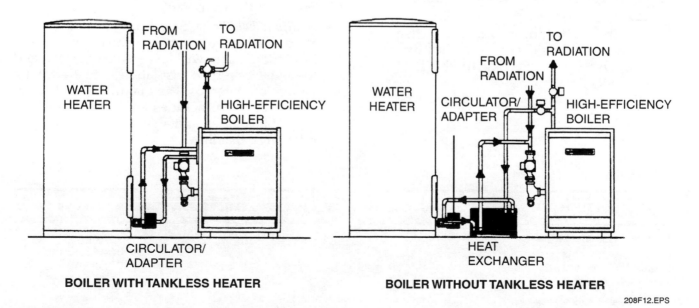

Figure 12 ◆ Indirect water heater with and without tankless heater.

Sections 2.0.0–4.0.0

1. Residential water heaters typically store the hottest water _____.
   a. in a special chamber of the tank
   b. mixed throughout the tank
   c. at the bottom of the tank
   d. at the top of the tank

2. The top element in most electric water heaters operates _____.
   a. whenever water is in the tank
   b. only when hot water demand is high
   c. only when hot water demand is low
   d. only when the lower element is not working

3. The _____ of a gas water heater transfer heat from burning gases to water in the tank.
   a. baffles
   b. thermocouples
   c. dip tubes
   d. thermostatic controls

4. The T/P relief valve of a gas water heater should be replaced if during testing it does not produce _____.
   a. a high limit reading
   b. a low limit reading
   c. water or steam
   d. compressed air

5. The high limit control of a water heater protects against extreme water temperatures by _____.
   a. shutting off the flow of water to the tank
   b. shutting off the flow of gas to the burner
   c. shutting off electricity to the heating elements
   d. releasing steam or water to reduce pressure in the tank

5.0.0 ◆ SELECTING WATER HEATERS

Gas water heaters that meet industry standards bear the **American Gas Association (AGA)** seal of approval. Electric water heaters that are tested and approved by the **Underwriters' Laboratories, Inc. (UL)** bear the UL seal.

When selecting the best water heater to fit individual installation needs and requirements, remember that no single unit or type is right for every situation. Consider the following when selecting a water heater:

- *Cost and availability of fuel* – In some areas, gas may not have been extended to the building. In some cases, electric heat may be too expensive.

- *Capacity* – The tank must be large enough to keep up with demand. Storage tank size and recovery rate determine water heater capacity. The **recovery rate** is the rate at which cold water can be heated. Recommended storage tank sizes and Btu (**British thermal unit**) ratings for water heaters are listed in *Table 1*. If more hot water is required, for example, to supply an automatic washer, use the next larger size storage tank or heater with a high recovery or Btu rating.

- *Durability* – A heater's ability to stand up to daily use depends on the material used to make it. Most units are made from galvanized steel, copper, or glass-lined steel. Most installers prefer glass-lined steel for its economy and durability. You can learn a lot about the tank's quality by reviewing the manufacturer's guarantee. A high-quality tank, properly cared for, should last 15 to 25 years or more. However, water conditions in different areas may shorten the life of the water tank.

- *Ability to hold heat* – Select fully insulated units to reduce heat loss and fuel consumption. The insulation is sandwiched between the tank and the outer covering. Insulation is made from a variety of materials. The manufacturer's product sheet often includes a description of the type of insulation and an indication of its heat-retaining capability.

Table 1 Sizing

Minimum Direct Gas-Fired Water Heater Sizes for One- and Two-Family Units										
Number of Bathrooms	1 to 1½			2 to 2½			3 to 3½			
Number of Bedrooms	2	3	4	3	4	5	3	4	5	6
Storage Tank Capacity (Gallons)	30	30	40	40	40	50	40	50	50	50
Input Rating (1,000 Btus per hour)	30	30	30	33	33	35	33	35	35	35
Source: Federal Housing Administration										

6.0.0 ◆ INSTALLING WATER HEATERS

The proper installation of a water heater takes place in two phases: (1) rough-in plumbing and (2) heater installation.

Before installing the water heater, inspect it for damage or defects. Measure the heater to ensure that there is ample space to set it properly. Most manufacturers provide a rough-in drawing showing the physical dimensions of the various parts of the heater as well as the specifications for capacity, input Btu per hour, and recovery rate (see *Figure 13*).

Be sure to supply proper venting if necessary. You may not need to vent gas-fired water heaters when outside air exchange is rapid. An example of this is large commercial buildings with many loading dock doors that open frequently or remain opened for long periods. Always check the local code. Electric water heaters have no flue gas and don't require venting. If you place gas or oil-fired heaters near a chimney or flue, as a rule, set these heaters no farther than 15 feet from such an opening to ensure a proper draft for safe operation.

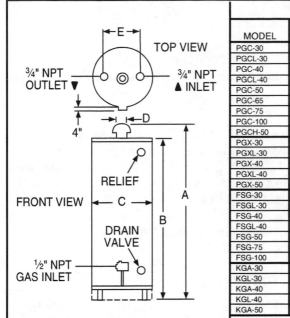

▲ INLET/OUTLET ON 75 GAL.=1"; ON 100 GAL.=1 1/4"

3/4" NPT RELIEF VALVE OPENING

		RECOVERY CAPACITIES				
		INPUT BTU/H GASES		RECOVERY: GPH 90°F RISE*		
MODEL	CAPACITY U.S. GALLONS	NATURAL	PROPANE	NATURAL	PROPANE	
PGC-30	30	30,000	30,000	30.7	30.7	
PGCL-30	30	30,000	27,000	30.3	27.2	
PGC-40	40	32,500	32,500	33.2	33.2	
PGCL-40	40	32,500	32,500	32.8	32.8	
PGC-50	50	40,000	40,000	40.9	40.9	
PGC-65	65	50,000	50,000	51.0	51.0	
PGC-75	75	80,000	75,500	67.2**	63.4**	
PGC-100	100	90,000	80,000	75.6**	67.2**	
PGCH-50	50	55,000	50,000	55.6	50.5	
PGX-30	30	30,000	30,000	30.3	30.3	
PGXL-30	30	30,000	27,000	30.3	27.2	
PGX-40	40	32,500	32,500	32.8	32.8	
PGXL-40	40	32,500	32,500	32.8	32.8	
PGX-50	50	40,000	40,000	40.4	40.4	
FSG-30	30	30,000	30,000	30.3	30.3	
FSGL-30	30	30,000	27,000	30.3	27.2	
FSG-40	40	32,500	32,500	32.8	32.8	
FSGL-40	40	32,500	32,500	32.8	32.8	
FSG-50	50	40,000	40,000	40.4	40.4	
FSG-75	75	75,000	60,000	75.7	60.6	
FSG-100	100	80,000	75,500	67.2**	63.4**	
KGA-30	30	40,000	37,000	37.7	34.8	
KGL-30	30	38,000	37,000	35.8	34.8	
KGA-40	40	40,000	37,000	37.7	34.8	
KGL-40	40	40,000	37,000	37.7	34.8	
KGA-50	50	40,000	40,000	37.7	37.7	

*RECOVERY CAPACITIES ARE BASED ON DOE METHOD OF TEST (90° RISE).
**RECOVERY CAPACITIES FOR MODELS OVER 75,000 BTU'S ARE BASED ON 100° RISE.
NOTE: TO COMPENSATE FOR THE EFFECTS OF HIGH ALTITUDE AREAS ABOVE 2,000 FEET, RECOVERY CAPACITY SHOULD BE REDUCED APPROXIMATELY 4% FOR EACH 1,000 FEET ABOVE SEA LEVEL.

10-YEAR	5-YEAR	KANKAKEE, IL						NEWARK, CA †					
		A	B	C	D	E	APPROX. WT. (LBS.)	A	B	C	D	E	APPROX. WT. (LBS.)
PGC-30		59	55 3/8	18 1/4	3	8	122	58 1/2	54 7/8	16 1/4	3	8	122
PGCL-30		46 3/4	43 1/8	20 1/4	3	8	117	46 3/4	43 1/8	20 1/4	3	8	117
PGC-40		57	53 3/8	20 1/4	3	8	137	57	53 3/8	18 3/4	3	8	137
PGCL-40		51 1/8	47 1/2	20 1/4	3	8	140	51 1/8	47 1/2	20 1/4	3	8	140
PGC-50		59 7/8	56 1/4	20 1/4	3	8	165	63 1/4	56 1/4	20 3/4	3	8	165
PGC-65		56 5/8	53 1/4	25 3/8	4	16	269	56 5/8	53 1/4	25 3/8	4	16	269
PGC-75		62 5/8	59 1/4	25 3/8	4	16	296	62 5/8	59 1/4	25 3/8	4	16	296
PGC-100		70 5/8	67	26 3/8	4	16	367	70 5/8	67	26 3/8	4	16	367
PGCH-50		64	58 1/4	20 3/16	4	8	165	64	58 1/4	20 3/16	4	8	165
PGX-30		59 1/2	55 7/8	15 3/4	3	8	110	59 3/8	55 3/4	16 1/4	3	8	107
PGXL-30		46 3/4	43 1/8	20 1/4	3	8	110	46 3/4	43 1/8	20 1/4	3	8	110
PGX-40		57 5/8	54	18 1/4	3	8	129	58	54 3/8	18 3/4	3	8	133
PGXL-40		51 1/8	47 1/2	20 1/4	3	8	136	51 1/8	47 1/2	20 1/4	3	8	136
PGX-50		60 3/4	57 1/8	20 1/4	3	8	161	64 1/8	57 1/8	20 3/4	3	8	154
	FSG-30	59 1/2	55 7/8	15 3/4	3	8	109	59 3/8	55 3/4	16 1/4	3	8	107
	FSGL-30	46 3/4	43 1/8	20 1/4	3	8	115	46 3/4	43 1/8	20 1/4	3	8	115
	FSG-40	57 5/8	54	18 1/4	3	8	125	58	54 3/8	18 3/4	3	8	133
	FSGL-40	51 1/8	47 1/2	20 1/4	3	8	140	51 1/8	47 1/2	20 1/4	3	8	140
	FSG-50	60 3/4	57 1/8	20 1/4	3	8	155	64 1/8	57 1/8	20 3/4	3	8	154
	FSG-75	61 1/8	57 3/4	25 3/8	4	16	257	61 1/8	57 3/4	25 3/8	4	16	257
	FSG-100	68 3/4	65 1/2	26 3/8	4	16	326	68 3/8	65 1/2	26 3/8	4	16	326
	KGA-30	59 1/2	55 7/8	15 3/4	3	8	110	59 3/8	55 3/4	16 1/4	3	8	107
	KGL-30	47 5/8	44	18 1/4	3	8	106	47 5/8	44	18 1/4	3	8	106
	KGA-40	57 5/8	54	18 1/4	3	8	124	58	54 3/8	18 3/4	3	8	133
	KGL-40	52 1/8	48 1/2	20 1/4	3	8	134	52 1/8	48 1/2	20 1/4	3	8	134
	KGA-50	64 5/8	57 1/8	20 1/4	3	8	157	59 7/8	56 1/4	20 3/4	3	8	154

† FOR DISTRIBUTION IN: ARIZONA, NEVADA, WASHINGTON, CALIFORNIA, OREGON, WESTERN MONTANA, IDAHO, UTAH, WESTERN WYOMING.

208F13.EPS

Figure 13 ◆ Rough-in sheet.

When you install connecting pipes from the water heater flue to the chimney, avoid using too many bends. Bends tend to block the movement of exhaust fumes through the chimney. Every bend decreases the length of run or footage. Make sure the flue connector is at least as large as the heater flue outlet.

If a furnace chimney is already installed, you may use it for the water heater as long as you connect the water heater flue to the chimney *above* the furnace flue. You must maintain a ¼-inch per foot upward slope in the horizontal run of the pipe. If you need to connect directly into a furnace flue, use a wye connector. Don't use a tee connector. The heat does not flow through a tee connector as efficiently as it does through a wye. Heat directed at an angle is more efficient.

Use galvanized or black-iron pipe and fittings for the gas or oil supply piping to the heating unit. Attach a shutoff valve to the piping between the water heater and the fuel supply. Anyone who replaces or repairs the heater can then turn this valve to shut off the fuel supply.

Use corrugated brass connectors to simplify hookup of the hot and cold water lines, but check to make sure their use is allowed by local codes. If the codes require regular pipe connectors, you must install **unions** (see *Figure 14*). Install a shut-off valve on the cold water inlet. Anyone who replaces or repairs the heater can then turn this valve to shut off the cold water supply.

Install the T/P relief valve in the specially designed opening in the top or side of the storage tank. To ensure safe operation of the water heater, follow the manufacturer's installation directions. Do not install plugs or valves between the T/P and the outlet discharge. Do not downsize the pipe between the T/P and the outlet discharge. Attach a drip line to the outlet of the T/P relief valve so that any hot water escaping from it will be directed safely to the sewer. Run the line so that it extends to within 6 to 12 inches of the floor drain, empties directly into a safety pan drain, or runs to the exterior of the building just above grade level. Check your local code to determine how to run the drip line. To prevent back siphoning, ensure that an air gap is maintained between the end of the drip line and the sewer. The air gap is maintained by leaving a space between the pipe and the top of the drain or standpipe.

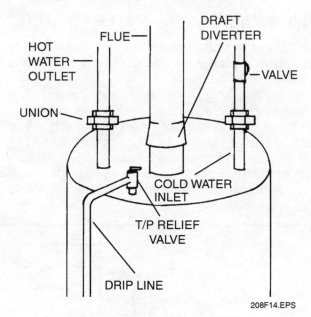

Figure 14 ◆ Use of unions.

When installing electric water heaters, note that they are manufactured in a range of voltage ratings—120-volt, 230-volt, or 440-volt. Make sure that the power source is the same voltage as the water heater. When disconnecting an old electric heater and installing a new heater, *never* assume that the breakers or fuses are labeled correctly. Always check the voltage with a voltage tester.

6.1.0 Parallel Connection

In some commercial installations with a high-volume demand for hot water, you can connect two hot water tanks in parallel (see *Figure 15*). The pipes leading from both feed lines to the cold water inlet and the hot water outlet must be equal in length. If they are not, the water will take the shortest path, and most, or all, of the hot water will be drawn from one tank. The result is that only one of the tanks will work, reducing the supply of unused hot water and increasing the chance that one of the heaters will fail early.

In a parallel connection, one heater can supply a building's hot water needs while the other is turned off for a brief time for maintenance, repair, or replacement. So long as the downtime for one heater occurs during a low-demand period, customers won't experience any decrease in water temperature.

ON THE LEVEL

Choosing Fixture Size

Be sure to match the fixture to the appliance. A typical 40-gallon water heater is not sufficient for large whirlpool tubs. This will leave the owner with a lukewarm bath.

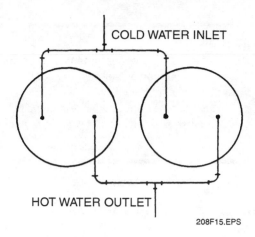

COLD WATER INLET

HOT WATER OUTLET

208F15.EPS

Figure 15 ◆ Parallel connection.

6.2.0 Series Connection

You can also connect water heaters in series. In a series connection, water is preheated in one tank to supply a second tank, which brings the water up to final temperature. The preheated water can also be supplied to the second tank by a coil output from an air-conditioner or by a solar source (see *Figure 16*).

DID YOU KNOW?

Solar Energy

The concept of solar energy really got cooking in 1767 when a Swiss scientist invented the world's first solar collector, or hot box. Sir John Herschel, a famous astronomer, cooked his food in a solar hot box during his expedition to Southern Africa in the 1830s.

   In the United States, solar water heating began when Clarence Kemp patented the first commercial water heater in 1891. The idea of using the sun to heat water was popular in areas that had to import fuel. By 1897, nearly 30 percent of the houses in Pasadena, California, had solar water heaters. In the 1970s, fuel shortages made solar water heating even more popular. The first commercial use of this technology appeared in the early 1980s. Today, utilities have more than 400 megawatts of solar thermal generating capacity.

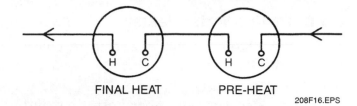

FINAL HEAT   PRE-HEAT

208F16.EPS

Figure 16 ◆ Series connection.

6.3.0 Non-Basement Installation

When you install a water heater in a location other than a basement, place a pan beneath the heater to collect any leakage. Run a 1-inch or ¾-inch pipe from the pan to the floor drain or waste opening. Pans come in different materials and sizes (see *Figure 17*).

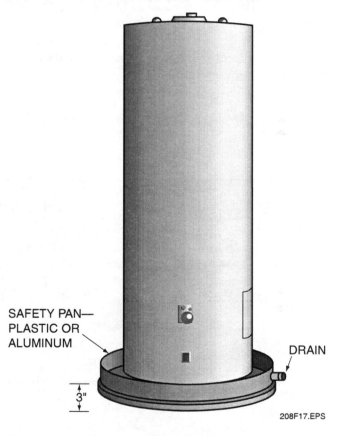

SAFETY PAN— PLASTIC OR ALUMINUM

DRAIN

3"

208F17.EPS

Figure 17 ◆ Safety pan and drain.

ON THE

·LEVEL·

Checking Water Heaters

Dirt and grime, small cracks, or damaged connectors can cause major problems once the water heater is installed and operating. Be sure to check each water heater for damage before installing it.

6.4.0 Testing Water Heaters

Testing water heaters should be a company policy for the following reasons:

- It eliminates callbacks.
- Callbacks cost the plumber both time and money.
- It prevents explosions from occurring (in gas-fired water heaters).

6.4.1 Testing Gas-Fired Water Heaters

After installing the water heater, do the following:

- Examine the heater to ensure that there are no leaks or drips.
- Apply a soap solution to joints and fittings and look for bubbles that would indicate a possible gas leak. (If you find a leak, examine the fittings for problems and replace them.)
- Light the pilot and turn the control valve up and down to ensure that the burner works.
- Turn the control knob to low and wait to see if the burner will cycle off.

 WARNING!

Always check the water heater temperature setting. A customer may be seriously scalded by water that's too hot.

At 125°F, it takes about 1½ to 2 minutes for hot water to cause scalding. At 155°F, it takes about 1 second.

6.4.2 Testing Electric Water Heaters

After installing an electric water heater, do the following:

- Turn on the water and check for leaks.
- Use a meter to ensure that the heater is set at the proper voltage.
- Set the thermostat to the desired setting.

 CAUTION

In spite of a manufacturer's quality controls, fittings can have defects. Be sure to examine fittings for holes, cracks, or poor casting. Be sure to read and follow the manufacturer's installation instructions.

ON THE · LEVEL ·

Water Heater Stands

Gas-fired hot water heaters draw air for combustion from about 2 inches above the floor. Sometimes water heaters are installed in locations where flammable liquids are stored (for example, in a garage where cans of gasoline or paint thinner may be stored). These liquids give off vapors that are heavier than air. The vapors sink, puddle near the floor, and are drawn into the hot water heater. Result—an explosion.

Manufacturers offer hot water heater stands that raise the heater about 18 inches off the floor. This raises the appliance above the area where flammable vapors puddle. The stand looks like a small table. Consult the local code about the use of water heater stands in your area.

Review Questions

Sections 5.0.0–6.4.2

1. When installing or replacing an electric water heater, it is always safest to _____.
   a. install a heater stand
   b. check the voltage with a voltage tester
   c. assume that breakers and fuses are labeled correctly
   d. turn on the heating elements before putting water into the tank

2. A gas water heater flue may connect to an existing furnace chimney only if it is _____.
   a. above the furnace flue
   b. below the furnace flue
   c. installed with a tee connector directly into the furnace flue
   d. installed with a wye connector branching down to the furnace flue

3. The drip line for a gas water heater should be installed so that it _____.
   a. recycles back into the water tank
   b. connects directly into the floor drain
   c. maintains an air gap to prevent back siphoning
   d. includes a plug between the T/P relief valve and the outlet discharge

4. After installing a gas water heater, a plumber should test for gas leaks by _____.
   a. setting the thermostat to high
   b. lighting the pilot and burner
   c. setting the thermostat to low
   d. applying a soap solution to the joints and fittings

5. After installing an electric water heater, a plumber should _____.
   a. set the thermostat as high as possible
   b. turn the control to low to see if it cycles off
   c. check that the heater is set at the proper voltage
   d. apply a soap solution to the joints and fittings

Summary

The water heaters most commonly installed in homes and commercial buildings are gas-fired or electric automatic storage tank heaters. These heat and store water at a thermostatically controlled temperature for delivery on demand. While there are some differences, the main components of both gas-fired and electric water heaters are the same. Gas-fired heaters require a piping system that delivers fuel to the burner unit and a venting system that removes combustion products from the building. The installation of a water heater is a two-stage operation: first, rough-in plumbing, and then installation of the heater. Checking the water heater for damage; verifying that it will fit in the space planned for it; supplying proper venting for gas heaters; hooking up the water lines; installing a T/P relief valve; and testing the unit are all steps that should be performed for a proper installation.

Trade Terms Introduced in This Module

American Gas Association (AGA): A membership organization that acts as a clearinghouse for gas energy information.

Anode: A metal rod that helps keep the tank from corroding because electrolysis eats away the anode instead of the tank; hence the term *sacrificial anode*.

Anti-siphon hole: An opening near the top end of the dip tube on a hot water heater that prevents hot water from siphoning out if an interruption occurs in the cold water supply.

Aquastat: A thermostat that regulates the temperature of hot water in a hot water boiler.

Automatic thermostat: The primary device for starting and stopping the flow of electricity to the heating elements.

Baffle: A plate that slows or changes the direction of the flow of air, air-gas mixtures, or flue gases.

British thermal unit: The amount of heat needed to raise the temperature of one pound of water by 1 degree Fahrenheit. Abbreviated Btu.

Dip tube: A device that prevents cold water from mixing with hot water; it delivers incoming cold water through the stored hot water to the bottom of the tank.

Dry fire: A term that refers to turning on an electric water heater with no water in the storage tank, causing the elements to burn out.

Flue: A heat-resistant enclosed passage in a chimney that carries away combustion products from a heat source to the outside.

High-limit control: A safety device that protects against extreme water temperatures caused by defective thermostats or grounded water heater elements.

Immersion element: Electrically heated elements that, when exposed to water, quickly and efficiently transfer heat.

Indirect water heater: A water heater in which a heat exchanger increases the water's temperature.

Insulation: A material that provides high resistance to heat flow.

Jacket: The outer shell of a water heater made of enamel baked on steel.

Main burner: The chamber in which the fuel gas mixes with air and is ignited. The combustion in the chamber heats the water in the tank to a preset temperature.

Pilot burner: See *safety pilot*.

Recovery rate: The rate at which cold water can be heated.

Relief valve opening: An opening on a hot water tank that provides access for installation of a temperature/pressure (T/P) relief valve.

Safety pilot: A device that ignites the gas at the main burner when turned on by the thermostat. Also called a *pilot burner*.

Scale: The crust on the inner surfaces of boilers, hot water heaters, and pipes formed by deposits of silica and other contaminants in the water.

Solenoid valve: A valve opened by a plunger controlled by an electrically energized coil.

Storage tank: The enclosed tank that stores heated water; it allows heated flue gases to cover the bottom of the tank before they enter the vertical flue. Commonly referred to as a *center-flue tank*.

Tank drain valve: A valve used to drain the tank and flush out sediment and scale.

Temperature/pressure (T/P) relief valve: A safety device installed in the storage tank or as close to the tank as possible. It prevents the tank from exploding should the thermostat fail to operate properly and the water overheat.

Thermocouple: A small electric generator made of two different metals joined firmly together. It produces a small electric current that holds the safety shutoff gas valve open.

Thermostatic control: A sensing device that turns the gas supply on to heat water in the storage tank, or off when the water reaches the preset temperature.

Top cover: The top outside cover on a water heater.

Underwriters' Laboratories, Inc. (UL): A nonprofit, nongovernment organization that classifies, tests, and inspects electric devices to ensure compliance with the National Electric Code.

Union: A pipe fitting that connects the ends of two pipes.

Valve control unit: In a water heater, the unit that contains the automatic pilot valve, or solenoid valve, and the thermostat valve.

Water heater control: In a water heater, the control that adjusts the water temperature.

Answers to Review Questions

Sections 2.0.0–4.0.0
1. d
2. b
3. a
4. c
5. c

Sections 5.0.0–6.4.2
1. b
2. a
3. c
4. d
5. c

ACKNOWLEDGMENTS

Figure Credits

A. O. Smith Company 208F01, 208F03, 208F04

Rheem Water Heater Company 208F09

Weil-McClain 208F11, 208F12

NCCER CRAFT TRAINING USER UPDATES

The NCCER makes every effort to keep these textbooks up-to-date and free of technical errors. We appreciate your help in this process. If you have an idea for improving this textbook, or if you find an error, a typographical mistake, or an inaccuracy in the NCCER's Craft Training textbooks, please write us, using this form or a photocopy. Be sure to include the exact module number, page number, a detailed description, and the correction, if applicable. Your input will be brought to the attention of the Technical Review Committee. Thank you for your assistance.

Instructors – If you found that additional materials were necessary in order to teach this module effectively, please let us know so that we may include them in the Equipment and Materials list in the Instructor's Guide.

Write: Curriculum Revision and Development Department
National Center for Construction Education and Research
P.O. Box 141104, Gainesville, FL 32614-1104

Fax: 352-334-0932

E-mail: curriculum@nccer.org

Craft _____ Module Name _____

Copyright Date _____ Module Number _____ Page Number(s) _____

Description _____

_____

_____

_____

_____

(Optional) Correction _____

_____

_____

(Optional) Your Name and Address _____

_____

_____

Fuel Gas Systems

COURSE MAP

This course map shows all of the modules in the second level of the Plumbing curriculum. The suggested training order begins at the bottom and proceeds up. Skill levels increase as you advance on the course map. The local Training Program Sponsor may adjust the training order.

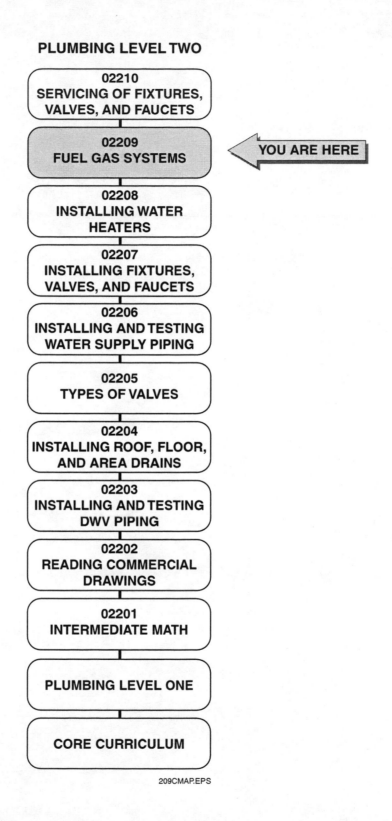

PLUMBING LEVEL TWO

02210
SERVICING OF FIXTURES, VALVES, AND FAUCETS

02209
FUEL GAS SYSTEMS ← YOU ARE HERE

02208
INSTALLING WATER HEATERS

02207
INSTALLING FIXTURES, VALVES, AND FAUCETS

02206
INSTALLING AND TESTING WATER SUPPLY PIPING

02205
TYPES OF VALVES

02204
INSTALLING ROOF, FLOOR, AND AREA DRAINS

02203
INSTALLING AND TESTING DWV PIPING

02202
READING COMMERCIAL DRAWINGS

02201
INTERMEDIATE MATH

PLUMBING LEVEL ONE

CORE CURRICULUM

209CMAP.EPS

Figures

Tables

Fuel Gas Systems

Objectives

When you have completed this module, you will be able to do the following:

1. Identify the major components of the following fuel systems and describe the function of each component:
   - Natural gas
   - LPG (liquefied petroleum gas)
   - Fuel oil
2. Identify the physical properties of each type of fuel.
3. Identify the safety precautions and potential hazards associated with each type of fuel and system.
4. Properly connect appliances to the fuel gas system.
5. Apply local codes to various fuel gas systems.
6. Design, size, purge, and test fuel gas systems.

Prerequisites

Before you begin, it is recommended that you successfully complete the following modules: Core Curriculum; Plumbing Level One; Plumbing Level Two, Modules 02201 through 02208.

Required Trainee Materials

1. Paper and sharpened pencil
2. Appropriate Personal Protective Equipment

1.0.0 ◆ INTRODUCTION

Fuel gases and fuel oil systems are a vital part of everyday life, providing energy to cook our food, heat our homes, and even power our vehicles. Traditionally, the plumber has been responsible for the safe distribution and use of fuel in buildings. As a plumber, you need to know about the characteristics of each type of fuel so that you will know how to deal with pipe sizing, piping materials, and the appliances that use each fuel type.

The raw materials used to produce fuel gases and fuel oil are called hydrocarbons, also known as petroleum. Petroleum requires considerable processing and refining before the raw materials can be converted from their original form into finished products such as gasoline, motor oil, and fuel oil.

Crude oil and natural gas, the hydrocarbons used to produce fuel gas and fuel oil, are found beneath the earth's surface. These rocks trap the petroleum deposits and prevent the oil and gas from dissipating or migrating into surrounding areas. These trapped deposits are called *pools*.

Petroleum consists of hydrogen and carbon atoms linked together to form chains of molecules. The basic hydrocarbon unit is the methane molecule, which consists of one atom of carbon (C) and four atoms of hydrogen (H).

$$\begin{array}{c} H \\ | \\ H - C - H \\ | \\ H \end{array}$$

This molecule may be combined with itself to form different chains of molecules, such as propane.

$$\begin{array}{ccc} H & H & H \\ | & | & | \\ H - C - C - C - H \\ | & | & | \\ H & H & H \end{array}$$

In general, short hydrocarbon chains are in the form of gases, whereas liquid petroleum is composed of longer, heavier chains:

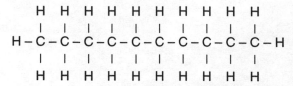

Crude oil is the natural form of oil as it is found in the earth. It is a chemically complex substance. Two of the fundamental classifications of crude oil are paraffin base (having a high wax content) and asphalt base (having a high asphalt content).

Crudes are also categorized according to their sulfur content. Sulfur has a distinct smell, so crudes are referred to as sweet (containing very little sulfur) or sour (containing a high amount of sulfur). The sweet crudes have less impact on air quality and on pollution problems than the sour crudes do.

Crude oil varies in color, texture, and density. Heavy crude oil may be so thick that it is solid at room temperature. This type of crude oil requires extensive refining before any useful product can be made from it. Light crude oil is lighter in weight and is easier to refine. Some crude oils need no processing at all.

Most crude oil comes from the earth mixed with foreign materials, such as water or particulates, that must be removed before the oil can be sold. Processed crude oil is converted into gasoline, fuel oil, lubricants, synthetics, and other petroleum products.

Natural gas refers to any of the gaseous hydrocarbons generated beneath the earth's surface. Natural gas is either dry or wet. Dry gas contains few or no liquid hydrocarbons. Wet gas contains liquid hydrocarbons, or **condensate.** Gasolines and other liquid products may be made from this natural gas condensate.

Although most natural gas does not require much processing, some gas must be dehydrated. This process removes any water and separates condensate from the gas. Natural gas also can be sour and may contain hydrogen sulfide. Hydrogen sulfide is a highly poisonous compound that must be removed before the gas can be used.

Mercaptans, compounds containing sulfur in place of oxygen, are usually added to natural gas to give the gas its familiar odor. This is done to make detection of gas leaks possible because natural gas is generally odorless. Propane and butane—forms of liquefied natural gas that are easily transported and stored—are often used for fuel in areas where other gaseous fuels are not readily available.

2.0.0 ◆ TYPES OF OIL AND GAS USED AS FUELS

As the *Introduction* illustrates, each fuel is composed of hydrocarbon chains. The length of these chains in each molecule of the gas or oil determines a variety of factors, such as the amount of Btus (a measure of thermal/heat units) given off when the fuel burns, how heavy the fuel is relative to air, and what safety considerations apply when it is used.

All the fuels have common installation elements: the materials used for piping and fittings, a plan outlining the size and location of pipes and tanks, the safe venting of the products of combustion, and consideration of the local codes that regulate each fuel's use. They also have some important differences. These common elements and differences are detailed in this module.

2.1.0 Natural Gas

Natural gas (methane) contains the basic hydrocarbon chain in each molecule. It is lighter than air, which means that natural gas tends to rise when released into the air. It has the lowest ratio of Btus per cubic foot of gas, which means that the pipes carrying natural gas at its normal pressure for use would be larger than the pipes used for other fuel gases or fuel oil.

Because natural gas is lighter than air, the model plumbing codes restrict its use under a building. If a leak occurred, natural gas would pool under a building's floors or floor slabs. Numerous gas explosions over the years would not have happened if modern plumbing codes had been observed at the time the piping was installed at those sites.

Natural gas is distributed by pipes from the gas well to the gas appliances that use it. Distribution of natural gas is regulated by the U.S. Department of Energy, the National Fire Protection Association, and the National Fuel Gas Code, as well as by state and local plumbing codes. These regulations describe the exact conditions that must be met and the equipment that must be used in distributing gas to consumers.

The difference between the various gas piping systems is the pressure of the gas, which ranges from thousands of pounds per square inch (psi) in the gas well to just less than 0.5 psi at the appliances. **Regulators** control the pressure at each stage from the gas choke on the wellhead to the final regulator at the gas meter.

Natural gas is sold by the cubic foot, and this is what the gas meter measures—cubic feet of gas used.

Specific Gravity

Specific gravity is a comparison of the weight of a gas or oil to an equal amount of air at a particular temperature, such as 60°F. Some gases such as methane (specific gravity = 0.55) are lighter than air (specific gravity = 1). Others, such as propane (specific gravity = 1.55), are heavier than air. Fuel oil is the heaviest of the fuels. Specific gravity is important because gases that are lighter than air will collect or pool at the top of a structure, and those that are heavier than air (propane) will collect or pool at the bottom of a structure. If a methane supply under a slab of concrete were to leak, the gas could collect and explode under the building. The same is true with heavier gas; if an appliance fueled by liquefied petroleum gas (LPG) is installed in a basement and a leak pools on the floor near a heater or a furnace, an explosion could occur. Often, local codes forbid the use of LPG appliances in basements.

Specific gravity also relates to the amount of Btus released in combustion. The heavier the gas or fuel oil, the more Btus it produces. This fact directly affects the size of the supply piping serving the appliance. Heavier gases and fuel oils require smaller pipes than do lighter gases. Plumbing codes recognize the differences in types of gases and fuel oils and their characteristics. Always consult the local code and manufacturers' specifications for working with fuel systems.

2.2.0 Liquefied Petroleum Gas (LPG)

LPG (liquefied petroleum gas), better known as LP or bottled gas, is an important source of fuel energy for the United States. It is most often produced from vapors during petroleum refining. The most commonly used LP gases are propane or a mixture of propane and butane.

LP gas is heavier than air and is colorless, odorless, and highly explosive. Bottled gas may be either a liquid or a vapor. Compressed under high pressure, the vapor may be changed into a liquid. This makes LP gas easy to store and transport. When the liquid is allowed to return to normal atmospheric pressure and temperature, it changes into a vapor. Bottled gas can be used as fuel only when it is in its vapor state. Because of its qualities, LP gas is a convenient source of fuel for rural areas or remote urban areas.

 DID YOU KNOW?
Propane Gas

Approximately 3 percent to 4 percent of the nation's total energy is supplied by propane gas. Residential and commercial sectors of the economy use some 6.8 billion gallons of propane a year. Industry uses about 8.2 billion gallons a year.

Because it is heavier than air, LPG tends to pocket in depressions when released into the air. For this reason, local codes often forbid the use of LPG appliances in basements.

LPG is sold by the gallon, and it is usually distributed by special tanker trucks that meter the gas flow and refill bottles or tanks at individual sites.

2.3.0 Fuel Oil

Fuel oil is a vital source of energy for residential and industrial use. It is transported and stored in a liquid state. However, it must be changed to a vapor or gas in order to be burned.

Fuel oils are most often identified by their **flash point** and **viscosity.** Flash point is the temperature at which fuel oil will ignite. Viscosity is a measure of the fuel's flowing quality or resistance to flow. Oil with a high viscosity is very thick and resists flowing. Oil with a low viscosity is thin and flows easily.

Fuel oil burner systems are designed to handle different grades or kinds of oil. The most common grades are No. 1 to No. 5. The higher the number, the higher the flash point and viscosity.

State and local codes regulate the materials that can be used in fuel oil installations.

Fuel oil is sold by the gallon, and it is distributed by tankers that meter the fuel oil as they fill the tanks.

1. _____ is lighter than air and has the lowest ratio of Btus per cubic foot of gas of the fuel gases.
   - a. Liquefied petroleum
   - b. Hydrogen
   - c. Methane
   - d. Oxygen

2. Because of the tendency of _____ to collect under floors or floor slabs in situations where a pipe might develop a leak, piping for this type of fuel is never installed underneath a building.
   - a. natural gas
   - b. propane
   - c. fuel oil
   - d. hydrogen

3. _____ control gas pressure all along the natural gas pipeline, reducing it from thousands of pounds per square inch at the source to just less than 0.5 psi at the appliance connection.
   - a. Gas meters
   - b. Gas chokes
   - c. Regulators
   - d. Pressure valves

4. Because of the tendency of _____ to collect in low spots or depressions, appliances using this type of fuel are never installed in a basement.
   - a. liquefied petroleum gas
   - b. natural gas
   - c. methane gas
   - d. fuel oil

5. Fuel oils are rated by both flash point and _____.
   - a. resistance
   - b. specific gravity
   - c. viscosity
   - d. compression

3.0.0 ◆ COMMON FACTORS IN FUEL SYSTEMS

This module covers three separate fuels—natural gas, LPG, and fuel oil—each of which has many different properties, uses, and installation procedures. However, the three also have many factors in common. These common factors are detailed in this section.

3.1.0 Materials

All three fuels are transported through pipe and piping systems.

3.1.1 Natural Gas Piping

Steel pipe (Schedule 40) is the most commonly used piping material in natural gas systems, although copper pipe is used in some circumstances, and plastic piping is being used more often.

Steel pipe is durable, versatile, and strong. It can be manufactured as **black iron** or as **galvanized pipe.** Black iron is less expensive because it is not coated with zinc. The carbon contained in steel gives black iron pipe its color. Black iron is used in installations where corrosion will not affect its uncoated surfaces. If black iron pipe is used in underground installations, it must be coated, wrapped, and cathodically protected (which means it must be made nonreactive to electricity).

Galvanized steel pipe is dipped in molten zinc, which gives the pipe a shiny, silver color. The zinc also protects the pipe's surfaces from corrosion and abrasive materials. Galvanized pipe may be installed above or below the ground.

Steel pipe joints may be screwed or welded. All screw fittings should be made of either malleable iron or steel.

Copper pipe (type K or L) can be used in gas system installations if the gas is not corrosive. Copper pipe can be connected by flared (if the joint is accessible), welded, or screwed joints. Flexible copper tubing with flared joints is usually used outside a building.

Polyethylene (PE) plastic pipe and fittings are gaining dominance in natural gas distribution. This flexible pipe is most often used underground outside a building. It is usually connected by mechanical joints. Be sure to consult the local gas code for the rules regulating the use of plastic pipe in your area.

Corrugated stainless steel tubing (CSST), which usually is covered with a plastic jacket, is used for gas piping. CSST systems are becoming more popular because of the lower labor costs involved in using flexible tubing that requires few joints. However, the tubing costs more than other materials and no single standard of manufacturing has been agreed on. Each CSST system is proprietary. This means that a manufacturer's pipe and fittings are suitable for use only with that manufacturer's system. Typically CSST systems are manifold systems; the gas is distributed from a large pipe with many openings to the appliances by CSST tubing with no other branches. Each run of pipe is a **home run** from the manifold to the appliance. The manifold design allows the manufacturer to make just a few sizes of piping; generally they are ⅜-inch, ½-inch, ¾-inch, and 1-inch nominal size. Naturally, these sizes limit this material to smaller commer-cial jobs and residential uses where larger pipe sizes would not be required. Special installation requirements are noted in manufacturers' instal-lation manuals. Many manufacturers require that installers be trained for certification in the use of their CSST systems.

Codes

Local building codes regulate and standardize piping installation and fuel storage. These codes set the minimum standard for safe use of these fuels, and you should always consult them before you design a fuel system. Together, the National Fire Protection Association (NFPA) and the American Gas Association (AGA) produce the National Fuel Gas Code. This code is used by many government agencies along with the model plumbing codes including the Building Officials and Code Administrators (BOCA), the Southern Building Code Confer-ence (SBCC), and the International Association of Plumbing and Mechanical Officials (IAPMO). These codes outline acceptable materials, fittings, valves, and installation pro-cedures and provide charts showing minimum pipe sizes for the fuel gases (see *Table 1*). Ask the local administrative authority for the name of the code you should consult.

Table 1 Size of Gas Piping

SIZE OF GAS PIPING
Maximum Delivery Capacity of Cubic Feet of Gas per Hour of IPS Pipe Carrying Natural Gas of 0.60 Specific Gravity Based on Pressure Drop of 0.5-Inch Water Column

Pipe Size, Inches	10	20	30	40	50	60	70	80	90	100	125
½	174	119	96	82	73	66	61	56	53	50	44
¾	363	249	200	171	152	138	127	118	111	104	93
1	684	470	377	323	286	259	239	222	208	197	174
1¼	1404	965	775	663	588	532	490	456	428	404	358
1½	2103	1445	1161	993	880	798	734	683	641	605	536
2	4050	2784	2235	1913	1696	1536	1413	1315	1234	1165	1033
2½	6455	4437	3563	3049	2703	2449	2253	2096	1966	1857	1646
3	11412	7843	6299	5391	4778	4329	3983	3705	3476	3284	2910
3½	16709	11484	9222	7893	6995	6338	5831	5425	5090	4808	4261
4	23277	15998	12847	10995	9745	8830	8123	7557	7091	6698	5936

Pipe Size, Inches	150	200	250	300	350	400	450	500	550	600	
½	40	34	30	28	25	24	22	21	20	19	
¾	84	72	64	58	53	49	46	44	42	40	
1	158	135	120	109	100	93	87	82	78	75	
1¼	324	278	246	223	205	191	179	169	161	153	
1½	486	416	369	334	307	286	268	253	241	230	
2	936	801	710	643	592	551	517	488	463	442	
2½	1492	1277	1131	1025	943	877	823	778	739	705	
3	2637	2257	2000	1812	1667	1551	1455	1375	1306	1246	
3½	3861	3304	2929	2654	2441	2271	2131	2013	1912	1824	
4	5378	4603	4080	3697	3401	3164	2968	2804	2663	2541	

Source: 1994 Uniform Plumbing Code (Roger Rotundo)

3.1.2 LPG Piping

LPG systems use steel or copper piping, depending on circumstances. For LPG systems, most codes require the following specifications for piping and tubing:

- Piping or tubing from a tank should have a working pressure equal to or greater than the working pressure of the tank. This prevents overpressurizing and possibly overloading the lines, which may result in a break or a leak.

- All piping on the low-pressure side of the regulator should have a minimum working pressure of not less than 125 psig (pounds per square inch gauge pressure).

- Steel, seamless copper, soft flexible copper (type K or L), or brass tubing may be used for piping LPG. Steel and copper pipes are used most often.

WARNING!

Never join copper used in gas piping by soldering. If a fire occurs, the heat could melt the joint and add fuel to the fire. Solder joints also sometimes deteriorate underground, which will cause a costly gas leak.

3.1.3 Fuel Oil Piping

In fuel oil systems, soft copper tubing (type K or L) is the most commonly used piping material. Most areas allow copper tubing to be connected only with flare joints. The size of copper tubing most often recommended in fuel oil codes is ⅜-inch O.D. (outside diameter).

The only type of steel pipe permitted in fuel oil piping systems is black iron.

3.2.0 Plans

Every installation should include plans and specifications of the elements of the fuel system. These are usually required in order to get a permit from the local jurisdiction. Even when a permit is not required, a sketch of the installation is a vital tool for the organized plumber.

3.2.1 Design and Sizing

Each plan should include the following elements (see *Figure 1*):

- Pressure of the gas being carried
- Location and Btu requirements of the fixtures being served

- Lengths and sizes of piping from the meter or tank to the appliances being served
- Location of regulators or pumps
- Type and size of the exhaust venting system
- Specific types and brands of the tanks, valves, regulators, and equipment being installed

Two critical factors in planning the pipe and vent installation are the location and elevation of the tanks in relation to the equipment.

The factors that come into play when designing gas systems include the following:

- The kind of gas being delivered
- The Btu rating of the gas being used
- The developed length of the run from the meter or regulator to the farthest equipment
- The Btu demand of the equipment to be served
- The type of piping being used (for example, CSST has its own sizing charts)

Let's design a simple natural gas system. Suppose there is a furnace located 60 feet from the meter. The furnace is rated at 160,000 Btu/h and requires low-pressure natural gas. The natural gas in this area has a Btu rating of 1,000 Btu per cubic foot/hour (CFH). (The Btu/CFH rating can range between 850 and 1,100 depending on the source well. Your local gas supplier can give you the Btu rating for your area.) Refer to *Table 1*. The pipe size is shown in the left-hand column, and the length in feet of the run is shown in the top row. The numbers in the remaining rows and columns are CFH. You need to convert the Btu/h to CFH. Do this by dividing the input Btu/h of the appliance by the Btu/h rating of the gas as follows:

$$160,000 \div 1,000 = 160 \text{ CFH/hour}$$

Run your finger along the top row until you find the column headed 60 (the number of feet from the appliance to the meter). Look down this row for 160 (the CFH/hour). In this case, you won't find it, so you must move to the next higher number, 259. Now follow across that row to the left until you reach the column headed *Pipe Size in Inches*. You'll see that a 1-inch pipe is required. Later in your training, you will learn how to size more complex piping systems.

3.3.0 Manufacturer's Installation Instructions

The petroleum industry is constantly changing. Equipment and appliances are always being improved to increase safety and efficiency. No one could keep track of all the changes, and that is why the manufacturer's instructions are so

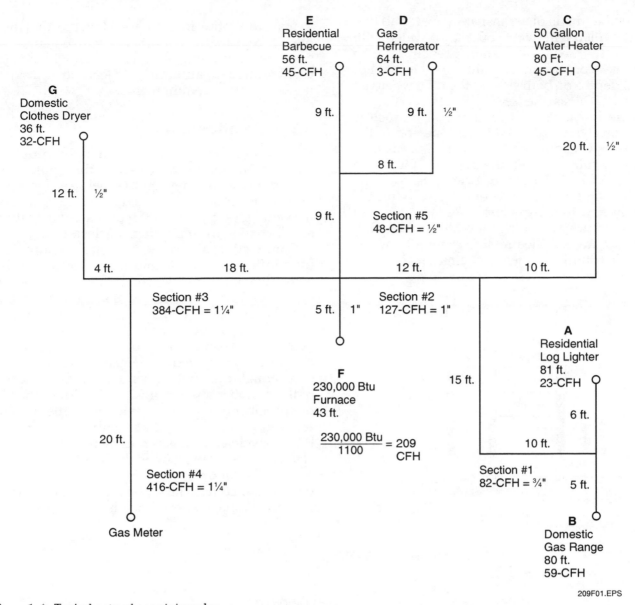

Figure 1 ◆ Typical natural gas piping plan.

important to the modern plumber. Each product or appliance comes with a set of instructions that outlines the clearances required, the special considerations on inlets and outlets, and the many other factors that the manufacturer requires for proper installation. Often, the manufacturer's warranty is based on correctly following these installation instructions. You should gather this information as each appliance is delivered and safeguard it for your use and for delivery to the customer after the installation is complete.

3.4.0 Testing

Service line installations from the curb to the structure are tested before the trench is backfilled. When you have completed the rough-in of a piping system, cap or plug all openings with a

threaded cap or plug and leave them closed until the appliance is connected. *Never* use wooden plugs, corks, or other improvised methods of closing pipe ends.

Next, test the roughed-in piping for leaks. It is important to do this test now, before walls and ceilings are finished, so that you won't have to cut holes in them to find and fix leaks.

To test for leaks, you can fill the piping with the fuel gas, air, or an inert gas such as helium, neon, nitrogen, or xenon. These inert gases are very stable and have extremely low combustion rates. *Never* use oxygen to test or **purge** gas lines.

To test for a leak, apply a soap-and-water solution to connections when the pipe is filled with air or inert gas. Bubbles indicate a leak.

Inside the building, a **manometer** is also used to test for leaks in the piping. The manometer is a U-shaped tube (made of a transparent material) that is usually filled with water, but may be filled with mercury to measure higher pressures. The manometer contains an amount of water that lies at equal depths on both sides of the tube when not connected to a gas line (see *Figure 2*).

A scale in inches and fractions of an inch is marked on the manometer. When one end of the manometer is connected to the gas line (usually by a small rubber hose), the pressure forces the water out of its balanced, or even, position. The water line on the other side of the tube, which is not connected to the gas line, rises. You can find the gas pressure by reading the scale. This reading is quite accurate. If the pressure gradually drops, the gas distribution system contains a leak.

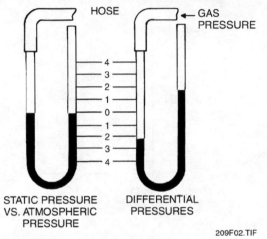

209F02.TIF

Figure 2 ◆ Manometer.

Instead of the manometer, local authorities often accept the use of other sensitive gauges to pressure-test piping systems. You should always check with these authorities to learn their inspection and testing requirements.

3.5.0 Combustion Air

Every fuel-burning appliance must be supplied with combustion air, which serves three purposes: (1) it provides the oxygen the fuel needs to support combustion, (2) it provides dilution air for the venting system, and (3) it provides ventilation air for cooling the equipment enclosure.

Combustion air can sometimes be supplied from infiltration of air into the space where the equipment is located. Whether this can be done is often based on the cubic feet of space in the room and the extent to which it is airtight. Modern buildings are often so tightly sealed that infiltration is not possible, and in those cases, ducting of outside air is required (see *Figure 3*).

Although the ducting is not often in the scope of the plumber's work, it would be irresponsible of a plumber to connect a device to the fuel system without determining that combustion air had been considered and was present in some form. A starved burner can result in the buildup of carbon monoxide, poor or nonexistent venting, and burner flameouts. These conditions can result in property damage or even death.

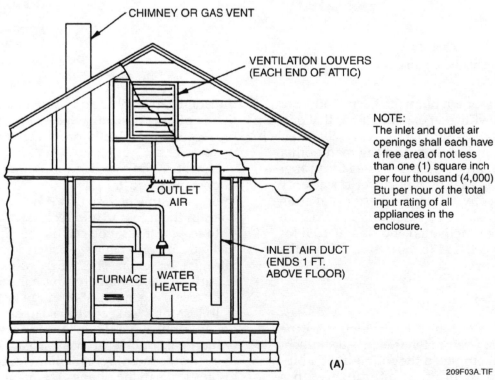

NOTE:
The inlet and outlet air openings shall each have a free area of not less than one (1) square inch per four thousand (4,000) Btu per hour of the total input rating of all appliances in the enclosure.

209F03A.TIF

Figure 3 ◆ Air supply methods.

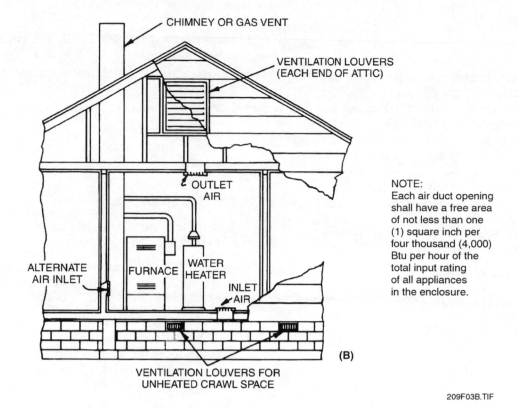

CHIMNEY OR GAS VENT

VENTILATION LOUVERS
(EACH END OF ATTIC)

OUTLET
AIR

ALTERNATE
AIR INLET

FURNACE

WATER
HEATER

INLET
AIR

NOTE:
Each air duct opening
shall have a free area
of not less than one
(1) square inch per
four thousand (4,000)
Btu per hour of the
total input rating
of all appliances
in the enclosure.

VENTILATION LOUVERS FOR
UNHEATED CRAWL SPACE

(B)

209F03B.TIF

CHIMNEY OR GAS VENT

OUTLET AIR DUCT

FURNACE

WATER
HEATER

INLET AIR DUCT

NOTE:
The inlet and outlet air
openings shall each have
a free area of not less than
one (1) square inch per
four thousand (4,000)
Btu per hour of the total
input rating of all
appliances in the
enclosure.

(C)

209F03C.TIF

Figure 3 ◆ Air supply methods.

3.6.0 Venting

Heating appliances that burn fuel must be vented to the exterior (see *Figure 4*). The combustion products of a fuel-burning appliance are dangerous. They can cause injury or death if they are not properly disposed of. The vent must not be smaller than the vent collar of the appliance.

Type B gas vents are used for the gas-fueled appliances (those using natural gas and LPG), and type L vents are used for oil-fueled appliances. Vents are part of an engineered system, and the manufacturer's instructions are vital for proper installation.

Examples of Type B gas fueled-appliances include the following:

- Central furnaces (warm-air types)
- Low-pressure boilers (hot water and steam)
- Water heaters
- Duct furnaces

- Unit heaters
- Vented room heaters (with appropriate input compensation)
- Floor furnaces (with appropriate input compensation)
- Conversion burners (with draft hoods)

Local codes give specific guidelines and restrictions for the installation of gas and oil venting, and must also be consulted.

3.7.0 Appliances

The four basic categories of gas-fueled appliances are described below.

- Category I: A noncondensing gas appliance that operates with a nonpositive vent pressure
- Category II: A condensing gas appliance that operates with a nonpositive vent pressure
- Category III: A noncondensing gas appliance that operates with a positive vent pressure
- Category IV: A condensing gas appliance that operates with a positive vent pressure

With Category I and III appliances, water will not collect internally or within the vent during continuous operation. Water may collect internally or within the vent during continuous operation of Category II and Category IV appliances. Nonpositive pressure means the vent is able to operate by natural, or gravity, draft. Positive pressure is introduced in a nonpositive system by adding a fan or burner that produces additional vent pressure to cause flow. Positive pressure is also found in systems where the internal static flue gas pressure is greater than the atmospheric pressure. In such systems, vent joints must be sealed to prevent leakage.

Some general installation practices are basic to safe appliance installation. Refer to the local code, the gas or oil supplier, and the manufacturer's instructions for more detailed information.

The authority having jurisdiction should approve all appliances, materials, equipment, and procedures used. In the case of LPG, for instance, they should be approved by the local gas code, by a nationally recognized authority such as the National Fuel Gas Code, or both. In general, approval is based on tests performed by nationally recognized testing laboratories such as the one operated by the AGA. These laboratories certify that the appliance, material, or equipment meets the minimum requirements of a national standard.

Determine whether the appliance is designed to operate on the gas to which it is to be connected. Never attempt to convert natural gas and LP gas appliances to burn fuel oil. All appliances have a

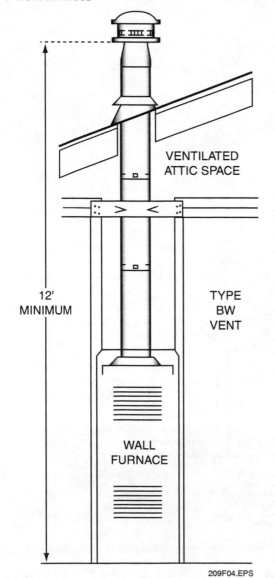

VENTILATED ATTIC SPACE

12' MINIMUM

TYPE BW VENT

WALL FURNACE

209F04.EPS

Figure 4 ◆ Typical vent for furnace.

manufacturer's label (see *Figure 5*) that provides critical information. Each label should detail the following:

- Type of gas required
- Btus the appliance uses (to ensure the proper size of supply pipe is used)
- Manufacturer's name
- Testing laboratory (Underwriter's Laboratory [UL], AGA) that lists the appliance
- Minimum clearance to combustible materials

Be sure to read and follow the manufacturer's installation procedures. All manufacturers' installation, operation, and maintenance instructions should be left at the job location.

Review Questions

Sections 3.0.0–3.7.0

1. Fuel oil supply piping lines are most commonly made using _____.
   a. galvanized steel
   b. black iron
   c. polyethylene
   d. type K or L copper

2. Copper tubing used for fuel gas supply piping lines for LPG should never be _____ because this would greatly increase the danger during a fire.
   a. flared
   b. solvent-welded
   c. soldered
   d. joined

> **CAUTION**
>
> All fuel oil and gas appliances must be installed away from flammable vapors and away from combustible materials.
>
> All appliances requiring venting must be vented according to the local code.
>
> All pipes must be adequately supported.
>
> All appliances must be located so that they are accessible for replacement or repair.

3. Helium, neon, nitrogen, and xenon are examples of _____ gases used for testing lines for leaks.
   a. inert
   b. pressurized
   c. manometer
   d. sensitized

4. A starved appliance will draw oxygen out of the surrounding atmosphere and produce lethal quantities of _____ if not ducted properly to provide combustion air.
   a. methane
   b. carbon dioxide
   c. carbon monoxide
   d. propane

5. _____ requirements to remove combustion products are determined by code based on the type of fuel used.
   a. Infiltration
   b. Venting
   c. Ducting
   d. Combustion

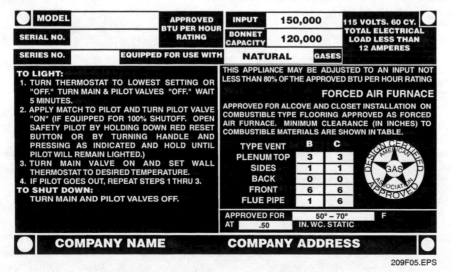

209F05.EPS

Figure 5 ◆ Appliance label.

4.0.0 ◆ FACTORS SPECIFIC TO GAS, LPG, AND FUEL OIL SYSTEMS

In addition to the factors described in the previous section that are common to gas, LPG, and fuel oil systems, many materials, installation and testing procedures, and other factors and requirements are specific to the type of fuel used. These are detailed in this section.

4.1.0 Natural-Gas-Specific Factors

Following are the materials, installation considerations, and other factors specific to natural gas. These factors include meters, valves, regulators, unions, **anodes,** protective coatings, and purging.

4.1.1 Meters

Gas meters are used to measure the amount of natural gas used by the consumer. Meters are located where they can be read easily and where their connections are readily accessible for service. Location, space requirements, dimensions, and the type of installation should be acceptable to the local gas supplier. Before locating the meter, you should contact the local gas supplier for specifications so that the meter can be located properly.

Most meters are classified as either tin case (see *Figure 6*) or aluminum case (see *Figure 7*). The outside casing is basically the only difference between the two types. The tin case meter, however, does have a stabilizer bar above the meter casing. This bar stabilizes the inlet and outlet connections. Gas companies are gradually replacing all tin case meters with aluminum case meters. Both kinds are positive displacement meters, meaning that no gas can pass through the diaphragm. The diaphragm moves in direct proportion to the amount of gas that flows. This measurement is recorded on the meter indexes in cubic feet (see *Figure 8*).

4.1.2 Valves

Valves are used to shut off gas flow. A flat-head cock valve with a lock wing (see *Figure 9*) is located before every meter. Other types of shut-off valves (see *Figure 10*) are positioned in front of the union

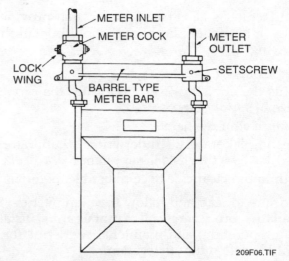

Figure 6 ◆ Tin case meter.

> **DID YOU KNOW?**
> All gas and oil is said to be flowing downstream from the well to its final use. When a valve is described as being located *after* a shut-off valve, this means that it is closer to its final destination—the appliance.

Figure 7 ◆ Aluminum case meter.

ON THE LEVEL

Appliance Connectors

Appliance connectors, which are often used on fuel gas systems, are flexible aluminum or stainless steel tubing that flares to screw pipe adapters on the end. They are installed after the appliance shut-off valve. Because the connectors are flexible, the appliance can be moved without being disconnected from the fuel system. These appliance connectors offer more resistance to flow than normal piping. Local codes restrict their lengths and location.

and are used to connect all gas appliances. A curb box (see *Figure 11*) with a **curb stop** is located between the gas main and the meter. Valves should be placed at every point where safety, convenience of operation, and maintenance demand.

4.1.3 Regulators

The regulator is located in or near the gas meter. It controls the amount of gas pressure from the meter into the building. It is self-regulating. If the gas pressure is low in the main service line, the regulator opens, allowing gas flow into the building. As the pressure increases in the service line, the regulator closes, restricting gas flow into the building.

4.1.4 Unions

An **insulating union** (see *Figure 12*) is always located on the gas main side of the meter. The union prevents the conduction of electricity through the customer's service line to the gas main. Other unions are required to connect appliances to the customer's gas distribution system.

These unions should be the ground joint (metal-to-metal) type. Unions make it possible to disconnect or remove an appliance. They are always located between the shut-off valve and the appliance. Concealed unions are prohibited.

Figure 9 ◆ Flat-head gas cock valve.

209F09.EPS

Figure 10 ◆ Level-handle gas cock valve.

209F10.EPS

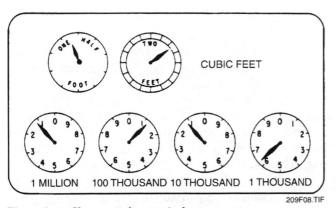

Figure 8 ◆ Close-up of meter index.

209F08.TIF

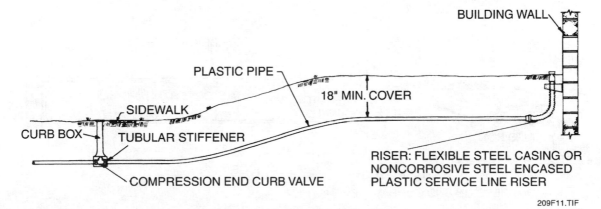

Figure 11 ◆ Curb box as part of system from gas main to meter.

209F11.TIF

Appliances and Appliance Fuels

Appliances are designed and built for specific fuels and are definitely not one-size-fits-all. Usually, the limiting factors are the orifices (mouth or vent openings) and the burner assemblies. The orifice and burner assembly for propane will be different from that built for natural gas, for instance. The gas companies usually have qualified personnel who can change out the orifice and then relabel the equipment for the gas being used. This should be done only by trained representatives of the gas company, and only as allowed by local code.

Figure 12 ◆ Insulating union.

209F12.EPS

4.1.5 Anodes

Anodes (see *Figure 13*) are used to ground the outside steel piping system. Any steel pipe or wrought iron pipe 2 inches or smaller in diameter that is 100 feet or less in length must have one 5-pound or greater magnesium anode. The anode should be buried a minimum distance of 2 feet from the pipe at or below trench depth. An anode should be located halfway between the curb box and the building.

If plastic pipe is used, each steel or malleable iron coupling must be individually protected by a zinc or magnesium anode. The anode is connected to the gas line or coupling by welding. Be sure to clean the anode wire connectors and the metal pipe or connector before you begin to weld. Both the anode and its connecting wire are to be coated with a proper coating material to protect the service line. Pipes over 100 feet in length require additional anodes. Consult your local code requirements.

4.1.6 Protective Coatings

All underground steel piping systems must be coated with a protective coating. This coating extends at least two inches on vertical sections of pipe above the **finished grade**. (Finished grade refers to the final landscaping.) The coating system consists of a primer and a coating that retards corrosion. Consult your local gas supplier for specifications and an approved list of coatings.

4.1.7 Purging

After the piping has been tested, all gas distribution systems must be purged. Purging removes all unwanted substances from the gas piping lines. One method of purging requires that the piping to an appliance be disconnected. Piping is never purged into the combustion chamber of an appliance. The open end of piping systems being purged should never be discharged into confined spaces or areas where there is a source of ignition.

After you have purged the gas piping, you must purge all appliances and light their pilots. Refer to appliance specifications and local code for appropriate methods.

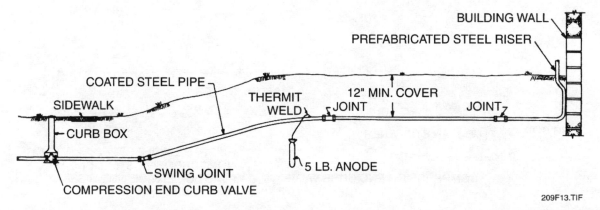

Figure 13 ◆ Anode as part of system from gas main to meter.

4.2.0 LPG-Specific Factors

Following are the materials, installation considerations, and other factors specific to LPG. These factors include storage containers, valves, regulators, vents, **suppressors,** gauges, unions, installation plans, and purging.

4.2.1 Storage Containers

LPG storage containers may be obtained according to the amount of gas required. When small amounts of gas are required for domestic use, cylinders are used. *Figure 14* shows some different types of LPG storage containers.

Capacities of large tanks are expressed in gallons. Cylinders, which can be lifted, are checked by weight (in pounds).

Most local codes require that LP gas storage containers be finished in a white or silver heat-reflecting surface, which reduces the amount of heat a container will otherwise absorb. This helps to prevent a buildup of pressure within the container.

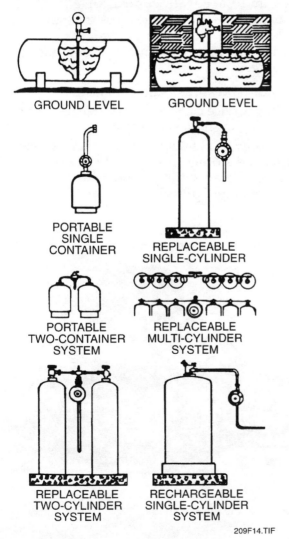

209F14.TIF

Figure 14 ◆ Different types of LPG storage containers.

4.2.2 Valves

A typical LP gas system uses a variety of valves. Because LP gas absorbs heat, it expands and increases its pressure on the inside of the container. You must install relief valves (see *Figure 15*) on all tanks or gas containers to prevent them from exploding when pressure increases. Relief valves are sized according to the needed discharge capacity of the container.

209F15.EPS

Figure 15 ◆ Relief valve.

CAUTION

You must have a plan of action in place before purging the lines so that you can perform this procedure safely. Some of the areas your plan should cover include the following:

• Controlling the purging rate
• Using proper ventilation
• Eliminating hazardous conditions

The mixture of gas and air released near the end of the purge is highly flammable. It must be dispersed before the appliance is ignited. Establishing a set waiting time before lighting the appliances is a good practice. Limit the amount of air being released so that the installer has plenty of time to react when just the fuel gas is coming out of the valve.

When there are concentrations of heavier-than-air fuel gases on the floor, or natural gas near the ceiling, fans may be needed to blow these concentrations outside. A hose temporarily connected to the end of the piping and run to the exterior may be an alternative to fans.

As on any work site, you must be aware of your working area and equipment and eliminate or correct conditions that can injure you or your co-workers. Always check with your supervisor if you have any questions about this or any other procedure.

Excess flow valves (see *Figure 16*) are designed to prevent the escape of gas when pipes or hose lines break. The valve actually shuts off the flow of gas if the flow becomes excessive. This prevents an entire tank from emptying into the atmosphere. Excess flow valves should have a rated closing flow (maximum allowed) that is 50 percent greater than the normal anticipated flow. The size of the pipe on either side of the flow valves should never exceed the size of the pipe of the excess flow valves. These valves should be tested at the time of installation and at least once a year after that.

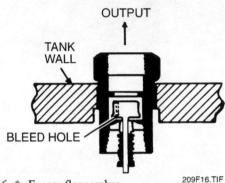

Figure 16 ◆ Excess flow valve.

209F16.TIF

Manual shut-off valves (see *Figure 17*) are required on bottled-gas containers. These valves must be located on the service outlet (building side) of the container. They must be accessible at all times.

209F17.EPS

Figure 17 ◆ Manual shut-off valve.

4.2.3 Regulators

Pressure regulators (see *Figure 18*) must be located on the outlet side (appliance side) of the LPG container. Regulators control the amount of gas pressure from the storage container to the appliances. Regulators are self-controlling. If the demand for gas is increased (if all the appliances are being used, for instance), the regulator allows greater gas flow into a building. As the need for gas decreases, the regulator restricts the gas flow.

Two regulators installed in the same service line increase the pressure control. The use of two regulators is known as *two-stage regulation*. The second regulator must be installed where the service line enters the building.

The first regulator reduces the pressure in the service line to 5–20 psig. It is referred to as the *high-pressure regulator*. The second regulator further reduces the flow of gas to a water column pressure of about 11 inches. (This is usually measured with a manometer.) The two-stage system gives more accurate control of gas pressure because of the reduced line pressure at the second regulator.

A two-stage regulation system has many advantages:

- Uniform pressure to all appliances
- Fewer service calls; regulator freeze-up is more common in single-stage systems
- Economy of installation; a smaller service line from the storage container to the building is needed because the second regulator reduces the amount of pressure

209F18.EPS

Figure 18 ◆ Pressure regulator.

PLUMBING LEVEL TWO — TRAINEE MODULE 02209

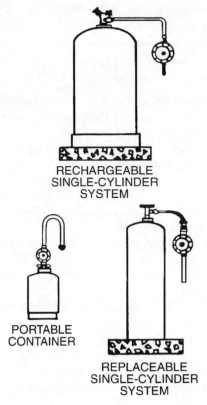

RECHARGEABLE
SINGLE-CYLINDER
SYSTEM

PORTABLE
CONTAINER

REPLACEABLE
SINGLE-CYLINDER
SYSTEM

209F19.TIF

Figure 19 ◆ Typical outside and inside regulator venting.

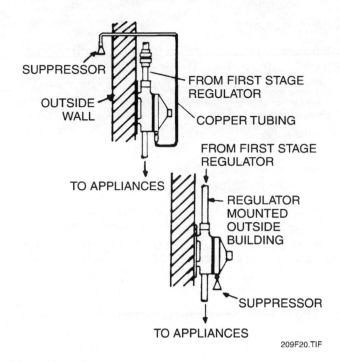

Figure 20 ◆ Suppressors.

4.2.4 Vents

All regulators must be vented (see *Figure 19*) so that the increased gas pressure caused by a malfunction in a regulator will be vented into the atmosphere and will not cause a service line to rupture. If a gas regulator is located inside, it should be vented outside.

4.2.5 Suppressors

Suppressors (see *Figure 20*) are placed on the regulator vent ends to retard the flow of gas. A rapid flow of gas into the atmosphere under high pressure could cause the vent outlet to freeze shut. A suppressor allows the gas to discharge at a rate that keeps the gas above freezing temperature.

4.2.6 Gauges

Gauges are used on large LP gas storage tanks to measure the amount of liquid in the container. A float gauge (see *Figure 21*) uses an indicator calibrated (marked) by percentage of total tank capacity.

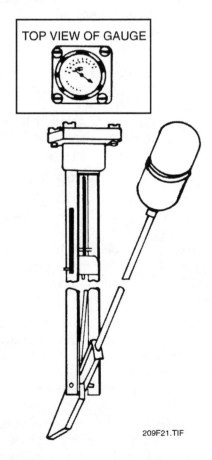

Figure 21 ◆ Float gauge.

4.2.7 Unions

A union is required to connect all appliances to the gas supply line. Unions make it possible to disconnect or remove an appliance. These unions should be of the ground joint (metal-to-metal) type. Concealed unions are prohibited. Unions must be located between the shut-off valve and the appliance.

4.2.8 Installation Plans

Residential plumbing drawings often do not include LP gas system instructions. Even if the gas supply lines are indicated, the local LP gas supplier must be consulted before the installation begins. The LP gas supplier will usually supply the following installation information:

- The size of the system required
- The location of the storage container
- The type of storage container needed
- What kind of regulator is needed and where it should be placed
- The customer service line requirements
- What inspections and tests are required and when they should be requested
- The appliance location and the type of venting system

After you get this information, mark it on a set of plans and use these plans as a reference during the gas system installation.

On building blueprints, gas supply lines can be identified by a solid line marked with the letter G.

4.2.9 Aboveground Installations

The plumber's responsibilities begin with sizing and locating the storage tank or cylinder. Consult your local LP gas supplier for specifications in your area. Storage tanks can be located above or below the ground. Here are some general specifications to follow when installing a tank above ground (see *Figure 22*).

- A horizontal tank must be mounted and secured on saddles (supports) that allow for expansion and contraction.
- The saddles must be large enough and positioned to distribute the container's weight equally.
- Corrosion-prevention measures must be taken to keep the portion of the container resting on the saddles or a foundation from rusting.
- Containers must be finished with a heat-reflecting surface (white or silver).

- Aboveground containers that have a 1,200-gallon (or greater) capacity must be electrically grounded.
- Aboveground containers must be placed well away from traffic areas.

Consult your local code and gas supplier for specific regulations governing aboveground storage tank installations in your area.

Usually, cylinders can be located according to the customer's preference and considering accessibility to service trucks. A typical cylinder location is shown in *Figure 23*. Remember to review local and state regulations and NFPA standards before locating the cylinders.

Aboveground storage containers cannot be used for all kinds of LP gas. For example, at 30°F and at atmospheric pressure, butane remains a liquid. So above ground containers are not suitable for butane which, like propane, only burns in vapor form. Butane was never used for household heating in the North and was sold for that purpose only in certain areas in the South. Because it does not make economic sense to distribute butane only to certain southern markets and propane to everyone else, the LP industry no longer distributes butane for home heating use. For retail household use and for most industrial uses, propane is the standard.

4.2.10 Underground Installations

These are some typical general specifications governing underground installation practices:

- Containers installed underground must be placed so that the top of the container is at least 2 feet below ground.
- Underground containers must be set on a firm foundation (earth). They can be surrounded by packed or tamped sand or soft earth if they are anchored with straps or cables.
- To prevent corrosion, an underground container must be primed and coated with a protective coating.
- In areas where soft soils and flooding are common, the tank must be secured to prevent floating.

Consult your local LP gas supplier for more specific requirements in your area, and locate storage tanks on the customer's property according to local specifications.

Regulators, service piping, and vents must be placed on LP gas systems according to local and state requirements. Exterior piping must be as direct as possible and must be installed at least 2 feet below the ground.

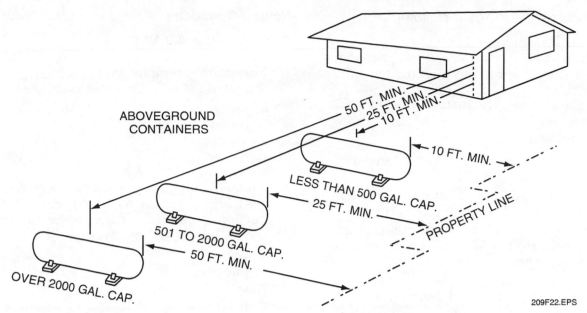

Figure 22 ◆ Storage tank locations.

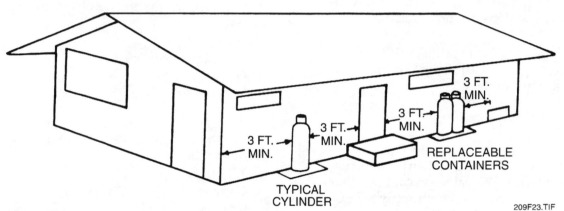

Figure 23 ◆ Cylinder locations.

Grade is very important to keep in mind when you are installing LP gas distribution systems. Remember that LP gas vapor will return to a liquid at a given pressure and temperature. Grade lines allow LP gas to flow back into the underground piping and change back into a vapor. **Drips** (see *Figure 24*) installed on underground lines allow storage of liquefied petroleum until it changes back to a vapor. Liquid LP gas flowing into a burner is very dangerous.

Drips must be constructed on horizontal and vertical runs of pipe or tubing. The drip pipe should be larger than the service line. A large drip pipe provides the needed capacity without using a long piece of pipe.

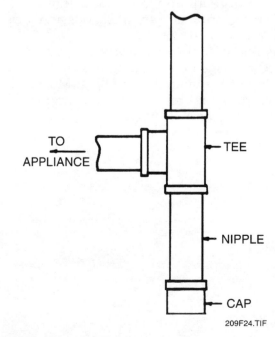

Figure 24 ◆ Drips.

4.2.11 Installations Inside a Building

LP gas piping inside any building must not be run in or through an air duct, clothes chute, chimney or flue, gas vent, ventilating duct, or masonry block.

LP gas regulations for installation inside a building are very specific. Typical regulations include the following:

- Piping and appliances must not be installed without sufficient ventilation.
- Piping under a building must have sufficient ventilation.
- When piping is installed aboveground, a drip should be buried underground to ensure vaporization of any liquid condensate.
- All LPG piping must be graded at least ¼ inch per 10 feet, pitched toward the storage tanks.
- Branch lines must be installed from the top or sides of lines. Whenever possible, 45-degree ells should be used on vertical gas piping. A 90-degree ell always offers the possibility of creating a trap.
- Where LPG piping is installed without the proper slope or grade, drips must be provided at the lowest point in the line.

4.2.12 Purging

After the piping has been tested, all LP gas distribution systems, like natural gas distribution systems, must be purged. The same precautions that apply to purging a natural gas system apply to LPG.

After you have properly purged the piping, you must purge all appliances and light their pilots. Refer to appliance specifications and local code for appropriate methods.

LP gas storage tanks must also be purged. Purging is accomplished by bleeding the air out of a vent on top of the tank during filling. Most local LP gas suppliers are responsible for purging the tanks when they refill them.

Review Questions

Sections 4.0.0–4.2.12

1. In a natural gas piping system, a(n) _____ valve is located before each meter along the line.
   a. regulator
   b. union
   c. anode
   d. flat-head cock

2. An insulating _____ is always located on the gas main side of the meter to prevent the conduction of electricity through the customer's service line to the gas main.
   a. valve
   b. purge
   c. union
   d. regulator

3. _____ absorbs heat, expanding and increasing its pressure on the inside of the storage container.
   a. Methane gas
   b. Liquid petroleum gas
   c. Hydrogen gas
   d. Natural gas

4. _____ regulate the flow of gas into the atmosphere to prevent the vent outlet of a regulator from freezing.
   a. Anodes
   b. Gauges
   c. Unions
   d. Suppressors

5. Proper _____ lines allow LP gas to flow back into underground piping and change back into a vapor.
   a. layout
   b. piping
   c. grade
   d. venting

4.3.0 Fuel-Oil-Specific Factors

Following are the materials, installation considerations, and other factors specific to fuel oil.

4.3.1 Filters

Filters are placed in the fuel oil line (**suction line**) between the storage tank and the burner to collect water and moisture that might be mixed with the fuel oil. A fuel filter must be located as close as possible to the burner.

4.3.2 Pumps

Pumps move the fuel oil through the piping system. They can push fuel, pull fuel, or both. A single-stage pump is used in **gravity feed** or low-lift systems. This kind of pump usually allows gravity to get the fuel to the pump, and then the pump pushes the oil into the burner. A single-stage pump should always be used on single-pipe or gravity feed systems (see *Figure 25*).

The two-stage pump pulls the fuel oil to the pump and then pushes it into the burner and returns the excess to the storage tanks located below the pump (see *Figure 26*). A two-stage pump should always be used in fuel piping systems that contain two or three pipelines.

4.3.3 Fuel Gauges

A fuel gauge is used to measure the capacity of the natural gas storage tank. A variety of fuel gauges are manufactured; use a type approved in your area.

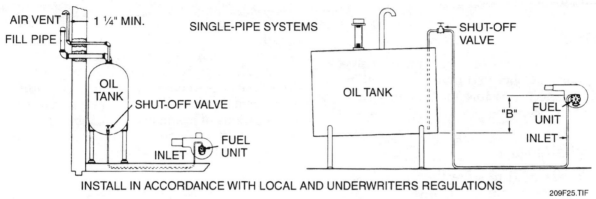

Figure 25 ◆ Gravity feed system.

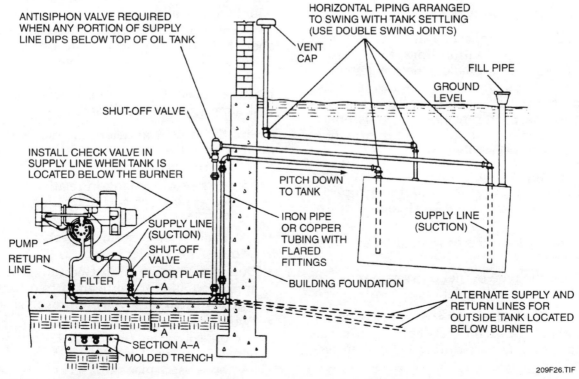

Figure 26 ◆ Two-stage pump system.

4.3.4 Valves

Valves are used to shut off the oil flow. A gate valve is usually located before (or in front of) the pump, on the return line, after the appliance, and between the suction line and the storage tank. Every appliance should have a shut-off valve on the fuel line before the union connection. Valves should be placed everywhere that safety, convenience of operation, and maintenance demand.

4.3.5 Vents

Vents are placed on all fuel oil storage tanks. They prevent excessive pressure from building up inside the tank. Most vents are constructed of galvanized iron pipe 1¼ inches in diameter. A vent cap is usually placed on a vent to prevent water from entering the tank.

4.3.6 Unions

Unions are required between the shut-off valve and the appliance. Concealed unions are prohibited. Unions make it possible to disconnect and remove an appliance.

4.3.7 Storage Tanks

Storage tanks are used to store the fuel oil (see *Figure 27*). These tanks are made of steel and are supported on concrete, masonry, or steel.

4.3.8 Installation Drawings

Residential plumbing drawings often do not include fuel oil distribution lines, but they do show the location of appliances. Before designing the fuel oil system, you should be aware of the local code regulations, the supplier's requirements, and the preferences of the customer. Gather and consider the following information before you start the installation:

- Determine the fuel oil storage tank location by consulting the supplier and the property owner. Keep in mind the local code requirements.
- Determine the locations of appliances and venting systems.
- Determine the type of fuel oil system required for each particular installation.

After you obtain the necessary information for the installation, mark it on a set of plans. Then, use these plans as a reference during the fuel oil system installation.

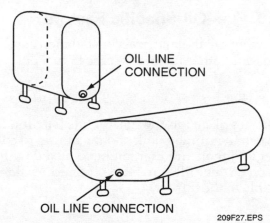

Figure 27 ◆ Typical storage tanks.

4.3.9 Layout

Layout means locating the exact position of pipes, fittings, appliances, and other fuel system materials. The layout is determined by information in the blueprints and codes. The fuel oil supplier often has specific requirements. Examples of layout work include the location of holes to be drilled, sleeves to be set, and supports. Other layout information includes size of pipe, type of pipe, and locations of branches. Remember that fittings restrict the flow of fuel, so design the system to use as few fittings as possible.

4.3.10 Aboveground Installations

Your responsibilities as a plumber begin with sizing and locating the fuel oil storage tank. Consult your local fuel oil supplier for the specifications in your area. Fuel oil storage tanks can be located above or below the ground or inside the building. They can also be enclosed or unenclosed.

The following are some general specifications that a plumber might encounter when installing a tank aboveground (see *Figure 28*):

- Size the system to determine the tank size. In any event, the tank cannot be larger than the code permits.
- Locate the storage tank in relation to the location of appliances and according to venting specifications.
- Locate tanks away from property lines or public traffic areas (refer to *Table 2*).
- Do not use more than two tanks of locally approved size or more than two tanks that together equal the locally approved size.
- Two tanks connected to the same burner may be cross-connected (see *Figure 29*) and provided with a single vent and fill opening. If this connection is used, place and secure both tanks on a common slab.

- Place a shut-off valve between the tank and the suction line.
- Size the vent pipe on a tank according to *Table 3*.
- Size and locate the fill opening to permit easy filling with minimal spillage.

Consult your local code and fuel gas supplier for specific information governing aboveground installations in your area.

4.3.11 Underground Installations

Many fuel storage containers are located below ground level. Location of fuel storage containers depends on the owner's preference. General specifications governing underground installation procedures include the following:

- Tanks installed underground must be covered with at least 2 feet of dirt.
- Underground containers must be set on a firm foundation (dirt). They must then be surrounded with clean sand, dirt, or gravel and must be well tamped in place.

- To prevent corrosion, tanks must be primed and coated with an approved protective coating.
- Tanks must be placed gently into the ground to prevent damage.
- Underground tanks must have an open or automatically operated vent.

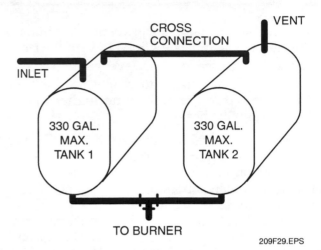

Figure 29 ◆ Cross-connected tanks.

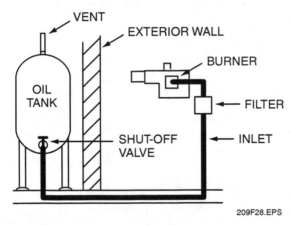

Figure 28 ◆ Typical aboveground system.

Table 3 Size of Vent Pipe on Fuel Oil Storage Tanks

Capacity of Tank, Gallons	Approximate Imperial Gallon	Diameter of Vent, Iron Pipe Size
500 or less	500 or less	1¼ in.
501 to 3,000	501 to 2,500	1½ in.
3,001 to 10,000	2,501 to 8,300	2 in.
10,001 to 20,000	8,301 to 16,600	2½ in.
20,001 to 35,000	16,601 to 29,000	3 in.

Note: Where tanks are filled by the use of a pump through tight connections, a vent pipe not less in size than the discharge of the pump shall be used.

Table 2 Location of Fuel Oil Storage Tanks

Capacity of Tank, Gallons	Minimum distance in feet from property line that is or can be built upon, including the opposite side of a public way	Minimum distance in feet from nearest side of any public way or from nearest important building on the same property
275 or less	5	5
276 to 750	10	5
751 to 12,000	15	5
12,001 to 30,000	20	5
30,001 to 50,000	30	10
50,001 to 100,000	50	15
100,001 to 500,000	80	25
500,001 to 1,000,000	100	35
1,000,001 to 2,000,000	135	45
2,000,001 to 3,000,000	165	55
3,000,000 +	175	60

ON THE

· LEVEL ·

LP License

In some states, a plumbing license does not cover LP installation. You'll need a separate LP license. Check the local code for information regarding licensing requirements for LP.

- When oil supply tanks are lower than the burner, the supply line must be sloped toward the tank.
- All tanks must have an oil-level gauge.
- Suction lines and fuel return lines must extend from the top of the tank to no higher than 4 inches from the bottom.

Consult your local fuel oil supplier for specific code requirements in your area.

The suction line from the tank to the building must be as direct as possible. Remember that fittings provide resistance to the flow of oil. Therefore, if possible, fittings should be eliminated from the suction line. Use valves, vents, fittings, and pipes that are approved by your local code. Oil lines entering a building through a masonry wall must be run through a sleeve (larger-sized pipe).

Pipes for oil distribution should not be smaller than ⅜-inch iron pipe or ⅜ inch (O.D.) copper tubing. The only exception occurs when the top of the tank is below the level of the fuel pump, in which case ¼ inch- or 5⁄16 inch-O.D. tubing may be used.

4.3.12 Installations Inside a Building

Fuel oil piping inside a building must not be run in or through an air duct, masonry block, or brick. If piping must enter a building through concrete, the pipe must be run through a sleeve, which should be encased in concrete. Fuel oil lines, especially those made of copper, must never be exposed to traffic areas. Avoid running fuel lines above the burner level because air has a tendency to collect at the highest point and stop the flow of fuel.

Fuel filters must be placed as close to the burner as possible. A ball check valve (see *Figure 30*) must

also be placed near the burner and in an accessible position. Codes in some areas suggest a valve on the return line to prevent siphoning from the storage tank into the burner. Care must be taken to open this valve before the burner is turned on to prevent the pump seals from rupturing.

Fuel oil regulations within a building are specific in most areas. Among the typical regulations are the following:

- Be sure the lines running to and from the burner do not prevent repairs to, or tests of, the burner.
- Be certain the fuel pump is the correct pump for your system.
- Design the pumping system according to the manufacturer's specifications.
- Protect vent and fill openings from the elements (rain and snow).
- Never use a single-line system on a buried tank.
- If possible, provide a loop of tubing in the suction line before the pump. This permits easier servicing and cuts down on vibration.

Fuel oil tanks may be installed within a building. This type of tank installation is specified as enclosed or unenclosed.

4.3.13 Enclosed Storage Tank Installations Inside a Building

General regulations for enclosed storage tanks (see *Figure 31*) include the following:

- The local codes will specify the maximum size of the tank that can be unenclosed, so any tank larger than that must be enclosed.

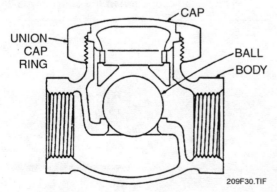

Figure 30 ◆ Ball check valve.

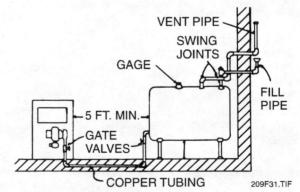

Figure 31 ◆ Enclosed tank.

WARNING!

Do not use an electric fan to ventilate an enclosed room. An explosion could result.

- Regardless of the enclosure, the size of a supply tank located above the lowest part (cellar or basement) of the building is limited by code. Check your local code.
- The floor, walls, and ceiling of the enclosure must have a fire resistance rating of not less than 3 hours. This means that it would take 3 hours before a fire could penetrate the walls.
- The tank must be supported at least 4 inches above the floor.
- Provisions must be made to properly ventilate the enclosed room.
- The enclosed tank must be properly vented to the outside atmosphere.
- The enclosed tank must be provided with a capacity-gauging device to allow measurement of fuel inside the tank.

4.3.14 Unenclosed Storage Tank Installations Inside a Building

Some tanks may be installed inside a building without being enclosed. Regulations concerning unenclosed tanks (see *Figure 32*) include the following:

- Your local code will advise you on the maximum size of tank that may be installed unenclosed, as well as other standards applying to the installation of unenclosed tanks inside a building.
- The supply tank must be of a size and shape that can be installed or removed from the building as a unit.
- The sizes of supply tanks that can be placed above the lowest part of the building must conform to local code standards.
- Unenclosed tanks cannot be closer than 2 feet to any source of heat.
- Unenclosed tanks must be supported and secured.
- A shut-off valve is to be placed between the tank and the supply line from the tank.
- The tank must be equipped with a vent and a fill gauge.
- Any unused opening on the tank must be plugged to prevent fuel oil vapors from escaping.

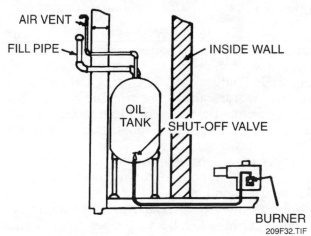

Figure 32 ◆ Unenclosed tank.

Review Questions

Sections 4.3.0–4.3.14

1. A _____ is required to move fuel oil through the piping system; it may be either single-stage or two-stage.
   a. filter
   b. pump
   c. valve
   d. union

2. _____ supported on concrete, masonry, or steel are used to store fuel oil.
   a. Distribution lines
   b. Canisters
   c. Tanks
   d. Saddles

3. Underground storage containers must have an open or automatically operated _____.
   a. valve
   b. vent
   c. regulator
   d. gauge

4. A _____ valve must be placed between the tank and the suction line.
   a. fill-opening
   b. one-stage
   c. spillage
   d. shut-off

5. For fuel oil storage tanks installed inside a building, the surrounding area must have a fire resistance rating of not less than _____ hours.
   a. 3
   b. 5
   c. 4
   d. 2

Summary

Fuel gases and fuel oil systems are an important part of everyday life. They provide energy for our homes and even power our vehicles. Each of the three fuels—natural gas, LPG, and fuel oil—share common factors and have specific differences. Plumbers must be aware of these factors so that they can properly plan, design, install, and test the systems and appliances that use these fuels. Plumbers are responsible for the safe distribution and use of fuel in buildings. Safe fuel gas and fuel oil piping systems are a result of proper installation techniques, quality materials, and strict adherence to code. Anything less than a proper piping installation can place the plumber, as well as the customer, in a potentially dangerous situation.

Trade Terms Introduced in This Module

Black iron: A grade of steel used in gas pipe installations.

Condensate: Water or other fluid that has been condensed from steam or from a vapor.

Curb stop (curb cock): A control valve that is placed in the gas supply line that runs between the gas main in a street to a building.

Drip: An extension of piping created by connecting a T, nipple, and cap that is installed on underground lines to allow storage of liquefied petroleum until it changes back into a vapor.

Finished grade: The surface or level of the ground.

Flash point: The temperature at which vapors given off by a fuel may be ignited.

Galvanized pipe: A pipe that has been coated with a thin layer of zinc.

Gravity feed: A system that depends on gravity to bring fuel to the pump. Gravity describes the tendency of objects or substances to move downward.

Home run: A run of pipe from a distribution point such as a manifold to the appliance with no branches. It serves only one appliance.

Insulating union: A nonconductive union placed on the gas main side of the meter to prevent electricity from being conducted through the customer's service line back to the gas main.

Layout: The location of the exact position of pipes, fittings, appliances, and other fuel system materials.

Manometer (draft gauge): A U-shaped tube containing fluid and marked with a graduated measuring scale, used for measuring slight changes in a low-pressure system such as a heating, venting, and air conditioning system. One end of the tube is connected to the low-pressure source. Slight changes in pressure cause the fluid in the tube to move up or down.

Purge: The removal of all unwanted substances from a gas piping line.

Regulator: A device designed to control flow or pressure.

Suction line: A line with negative pressure that will draw a liquid or gas.

Suppressor: A device placed on regulator vent ends to allow the gas to discharge at a rate that keeps the gas above freezing temperature.

Viscosity: A measure of a liquid's resistance to flow. The higher the viscosity, the thicker and slower a liquid flows.

Answers to Review Questions

Sections 2.0.0–2.3.0
1. c
2. a
3. c
4. a
5. c

Sections 3.0.0–3.7.0
1. d
2. c
3. a
4. c
5. b

Sections 4.0.0–4.2.12
1. d
2. c
3. b
4. d
5. c

Sections 4.3.0–4.3.14
1. b
2. c
3. b
4. d
5. a

ACKNOWLEDGMENTS

Figure Credits

National Fire Protection Association	209F22, 209F23
Ohio Brass	209F20
Sandstrand	209F16, 209F17, 209F18
Southern Building Code	209F03A, 209F03B, 209F03C
Westwater	209F15

NCCER CRAFT TRAINING USER UPDATES

The NCCER makes every effort to keep these textbooks up-to-date and free of technical errors. We appreciate your help in this process. If you have an idea for improving this textbook, or if you find an error, a typographical mistake, or an inaccuracy in the NCCER's Craft Training textbooks, please write us, using this form or a photocopy. Be sure to include the exact module number, page number, a detailed description, and the correction, if applicable. Your input will be brought to the attention of the Technical Review Committee. Thank you for your assistance.

Instructors – If you found that additional materials were necessary in order to teach this module effectively, please let us know so that we may include them in the Equipment and Materials list in the Instructor's Guide.

Write: Curriculum Revision and Development Department
National Center for Construction Education and Research
P.O. Box 141104, Gainesville, FL 32614-1104

Fax: 352-334-0932

E-mail: curriculum@nccer.org

Craft _____ Module Name _____

Copyright Date _____ Module Number _____ Page Number(s) _____

Description _____

_____

_____

_____

(Optional) Correction _____

_____

_____

(Optional) Your Name and Address _____

_____

_____

Servicing of Fixtures, Valves, and Faucets

COURSE MAP

This course map shows all of the modules in the second level of the Plumbing curriculum. The suggested training order begins at the bottom and proceeds up. Skill levels increase as you advance on the course map. The local Training Program Sponsor may adjust the training order.

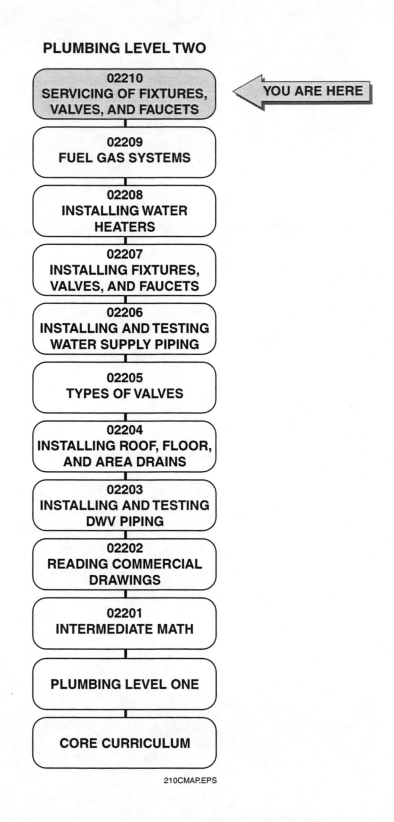

PLUMBING LEVEL TWO

02210
SERVICING OF FIXTURES, VALVES, AND FAUCETS ⬅ YOU ARE HERE

02209
FUEL GAS SYSTEMS

02208
INSTALLING WATER HEATERS

02207
INSTALLING FIXTURES, VALVES, AND FAUCETS

02206
INSTALLING AND TESTING WATER SUPPLY PIPING

02205
TYPES OF VALVES

02204
INSTALLING ROOF, FLOOR, AND AREA DRAINS

02203
INSTALLING AND TESTING DWV PIPING

02202
READING COMMERCIAL DRAWINGS

02201
INTERMEDIATE MATH

PLUMBING LEVEL ONE

CORE CURRICULUM

210CMAP.EPS

MODULE 02210 CONTENTS

Figures

Tables

Servicing of Fixtures, Valves, and Faucets

Objectives

When you have completed this module, you will be able to do the following:

1. Identify common repair and maintenance requirements for fixtures, valves, and faucets.
2. Identify the proper procedures for repairing and maintaining fixtures, valves, and faucets.

Prerequisites

Before you begin, it is recommended that you successfully complete the following modules: Core Curriculum; Plumbing Level One; Plumbing Level Two, Modules 02201 through 02209.

1.0.0 ◆ INTRODUCTION

Valves and faucets control the flow of fluids and gases. They serve an important function, and, as a plumber, you must know how to service and repair them. In this module you'll learn trouble shooting procedures that focus on the internal workings of valves, and you'll learn how to correct problems when they occur.

2.0.0 ◆ SERVICING OF FIXTURES, VALVES, AND FAUCETS

Before you can begin any repair work, there are some general safety guidelines you should know and follow. It is impossible to predict what you will encounter when responding to a service call. For example, a small leak from a faucet spout into a kitchen sink, while wasteful and annoying, is not an emergency. Usually it's also not hazardous. On the other hand, a valve leaking a lot of water above a suspended ceiling is usually hazardous. In an emergency like this, you must stop the flow of water, immediately minimize the potential damage, and make the area safe before you make the repairs.

Generally, repairs to fixtures involve a stoppage in the drain fixture or a cracked or leaking fixture. If a leak in a fixture is the result of a problem with the valve or faucet, you will usually repair or replace the valve or faucet. If the problem is with the fixture itself, you will replace the fixture.

2.1.0 General Safety Guidelines for Service Calls

Some general safety guidelines for you to follow when responding to a service call appear below. Adapt these guidelines to the job at hand. As you progress in your career, you'll add your own guidelines from your experience.

- Wear rubber-soled shoes or boots for protection from slipping and electric shock.
- Turn off electrical circuits.
- Shut off a valve upstream from the leak. (If you think it will take you awhile to locate and turn off this valve, direct the leak into a suitably sized container to minimize damage until you can turn the water off.)
- Remove excess water.
- Move furniture, equipment, or any other obstacles clear of the work area.
- Cover furniture, floors, and equipment to protect them from any damage that could occur as a result of your work on the valve.
- Place ladders carefully, bracing them if necessary.
- When finished, turn the water back on to test your repair.
- Do your work neatly and clean up when you are completely done with the job.

3.0.0 ◆ TYPES OF VALVES

Once you take the necessary safety precautions, you are ready to inspect and repair the valve. While valves and faucets come in many styles, they can be divided into eight categories for repair work purposes:

- **Globe valves**
- **Gate valves**
- **Flushometers**
- **Ball cocks**
- **Tank flush valves**
- **Cartridge faucets**
- **Rotating ball faucets**
- **Ceramic disc faucets**

3.1.0 Globe Valves

Internally, globe valves, angle valves, and compression faucets are all designed with the same basic parts (see *Figure 1*). Therefore, the types of problems and their solutions are similar. These problems and their likely causes are presented in *Table 1*.

DID YOU KNOW?
American Standard

American Standard is the world's largest producer of bathroom and kitchen fixtures and fittings. It is also a leading producer of air-conditioning and heating systems.

American Standard resulted when the American Radiator Company merged with the Standard Sanitary Manufacturing Company in 1929. At first, the new company was called the American Radiator & Standard Sanitary Corporation. In 1967, it changed its name to American Standard.

Standard Sanitary was formed in 1899. This company pioneered improvements such as the one-piece toilet, built-in tubs, combination faucets (which mix hot and cold water), and brass fittings that don't tarnish or corrode.

Table 1 Troubleshooting Globe and Angle Valves and Compression Faucets

Problems	Possible Causes
1. Drip or stream of water flows when valve is closed	Worn or damaged seat Worn or damaged seat disc
2. Leak around stem or from under knob	Loose packing nut Defective packing Worn stem
3. Rattle when valve is open and water is flowing	Loose seat disc Worn threads on stem
4. Difficult or impossible to turn handwheel or knob	Packing nut too tight Damaged threads on stem

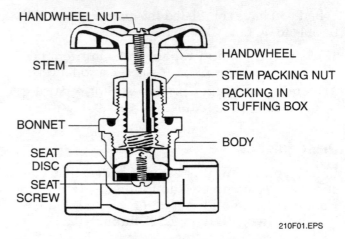

Figure 1 ◆ Basic parts of globe and angle valves and compression faucets.

When repairing valves, *always* use the correct size wrenches, and make sure their jaws are clean and smooth to prevent scratching the finished surfaces. Do not use pliers or adjustable wrenches.

If you have problem 1 or 3 in *Table 1*, remove the **valve bonnet** with a correctly sized wrench. Once you've removed the bonnet and stem assembly, inspect the seat disc and the seat. You can solve most cases of these two problems by replacing the disc and resurfacing or replacing the seat.

ON THE · LEVEL ·

Cleaning Up is Part of the Job

Whether you're installing or repairing valves, you must always clean up your work site. A clean work site is safer for you and for any trades that come in after you. Cleaning up becomes especially important if you are called in to repair a major leak. No one wants to pay to repair a valve or a faucet and then have to replace furnishings or carpet because you were careless.

A screw holds the seat disc in place. If you can't remove the screw or if the screw breaks off, replace the stem or use a drill and **screw extractor** (see *Figure 2*) to remove the broken screw. Once you have removed the old screw use a **tap** to recut or clean out the threads.

To replace the disc, select the correct disc and secure it to the end of the faucet stem. *Figure 3* illustrates the variety of disc (faucet washer) sizes available. Plumbers generally carry an assortment of sizes so that they are prepared to handle most repair jobs.

Replace the screw that holds the disc in place. Use screws with a soft metal or plastic plug near the tip. This plug locks the screw into position, eliminating the possibility that it will loosen.

To repair a damaged valve seat, fit a cutter into a **reseating tool** and insert it into the valve or faucet body (see *Figure 4*). When you turn the handle, the cutter refaces the valve seat. After a few turns of the handle, inspect the seat. If it is smooth, stop. If it is not smooth, give the handle another turn or two and inspect the seat again.

Some compression valves are equipped with replaceable seats. Simply replace the seat when it becomes worn or damaged. *Figure 5* shows a few of the many **removable seat wrenches** available. *Figure 6* shows a few of the many replaceable seats available. Most plumbers carry an assortment of seats in their trucks.

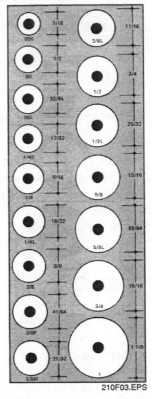

Figure 3 ◆ Faucet washer sizes (not to scale).

Figure 2 ◆ Screw extractor, drill bit, and screw.

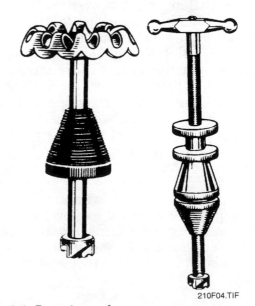

Figure 4 ◆ Reseating tools.

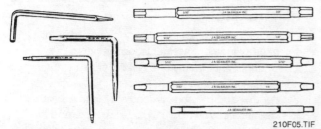

Figure 5 ◆ Removable seat wrenches.

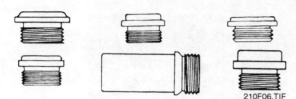

Figure 6 ◆ Replaceable seats.

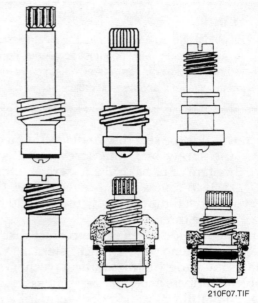

Figure 7 ◆ Stems.

Before you reinstall the stem, check the amount of play in the threads that join the stem to the bonnet. If these threads are badly worn, the stem may rattle when the valve is opened. Replacing the stem may solve the problem. Because many different types of stems are available, you'll need to know the name of the valve manufacturer to get the right replacement. You may have to take the old stem to the supplier for a direct comparison to the replacement part. *Figure 7* shows a few of the many types of stems available. Note that if the threads in the valve body are badly worn, you should probably replace the entire valve.

Refer to *Table 1*. To solve problems 2 and 4, remove the **packing nut** and inspect the packing. Leaks around the stem generally result from wear of the packing or the stem shaft. If the stem shaft is excessively worn, replace the stem. If the problem is the packing, replacement of the packing and lubrication should solve the problem. Use a **packing extractor** to remove the packing (see *Figure 8*). Before you select the replacement packing, examine the size and shape of the area under the packing nut. *Figure 9* illustrates some of the various sizes and shapes of pre-formed packing available. If the correct size and shape of pre-formed packing is not available, you can use twist packing (see *Figure 10*). Install the **twist packing** by wrapping the string around the stem to fill the packing area. When you tighten the packing nut, the packing forms into the required shape. Twist packing is not as durable as the pre-formed packing, which is made specifically to fit the valve.

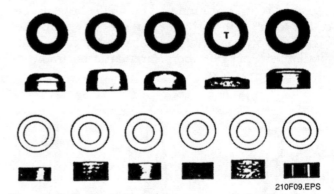

Figure 8 ◆ Packing extractors.

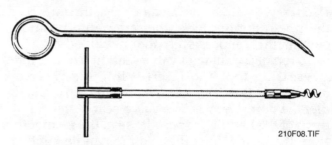

Figure 9 ◆ Pre-formed packing.

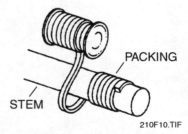

PACKING

STEM

Figure 10 ◆ Twist packing.

10.4

If the stem is difficult to turn, the reason may be one of the following:

- The packing nut is too tight. Loosening the nut and lubricating the stem where it passes through the packing should solve the problem. A special lubricant that resists high temperature is available. In addition to being heat resistant, it also makes the joint more waterproof.
- The threads on the stem are damaged. If the threads in the base of the valve are not damaged, replacing the stem should solve this problem. If these threads are damaged, replace the valve.

3.2.0 Gate Valves

Some of the problems common to globe valves are also common to gate valves (see *Figure 11*). A leak around the stem is the result of either packing wear or wear on the stem. Use the same procedure described for globe valves to solve this problem.

Gate valves are designed to be either fully opened or fully closed. They are not intended to throttle the volume of liquid flowing through the pipe. When gate valves are opened only part way, the disc (gate) has a tendency to vibrate, which causes the edge to erode. This defect is commonly called **wire drawing**. Repair of the valve is not likely to solve this problem. More than likely you will have to install a ball valve or a globe valve.

If a gate valve fails to stop the flow of water, one of three possible problems may be the cause:

- The disc (gate) is worn.
- The valve seat is worn.
- Some foreign material is preventing the disc from seating.

To determine which problem exists, remove the bonnet and inspect the gate and seat. You may be able to solve the problem by scraping out mineral deposits and carefully cleaning the mating surfaces. You may have to replace the disc, and if the seat area is worn, you'll have to replace the valve. Note that some companies remanufacture larger gate valves. Gate valves that are 2 inches in diameter and larger may be exchanged for remanufactured gate valves.

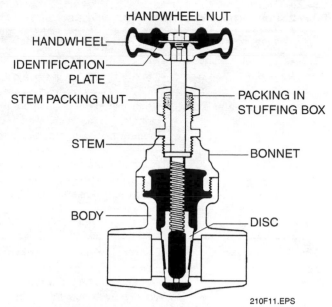

Figure 11 ◆ Gate valve.

3.3.0 Flushometers

The troubleshooting guide in *Table 2* applies to flushometers (see *Figure 12*). Four common problems are associated with flushometers:

- Leakage around the handle
- Failure of the vacuum breaker
- Malfunction of the **control stop**
- Leakage in the diaphragm that separates the upper and lower chambers

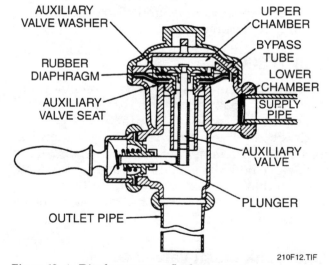

Figure 12 ◆ Diaphragm type flushometer.

ON THE · LEVEL ·

Repair or Replace?

Often replacing a defective part makes more sense than trying to repair it. It can be difficult to find the right parts. You can save yourself and your customer a lot of time and money by simply replacing defective parts. Every situation is going to be a little different, so be sure to think about cost effectiveness and customer service when deciding whether to repair or replace a part.

Table 2 Troubleshooting Guide for Flushometers

Problem	Cause	Solution
1. Nonfunctioning valve	Control stop or main valve closed	Open control stop or main valve
2. Not enough water	Control stop not open enough Urinal valve parts installed in closet parts Inadequate volume or pressure	Adjust control stop to siphon fixture Replace with proper valve Increase pressure at supply
3. Valve closes off	Ruptured or damaged diaphragm	Replace parts immediately
4. Short flushing	Diaphragm assembly and guide not hand-tight	Tighten
5. Long flushing	Relief valve not seating	Disassemble parts and clean
6. Water splashes	Too much water is coming out of faucet	Throttle down control stop
7. Noisy flush	Control stop needs adjustment Valve may not contain quiet feature The water closet may be the problem	Adjust control stop Install parts from kit Place cardboard under toilet seat to separate bowl noise from valve noise—if noisy, replace water closet
8. Leaking at handle	Worn packing Handle gasket may be missing Dried-out seal	Replace assembly Replace Replace

Kits that contain the components you'll need to repair each of these defects are available. Even though all flushometers contain the same basic components, specific parts will vary from manufacturer to manufacturer. *Always* follow the manufacturer's specifications and instructions when installing replacement parts.

3.4.0 Ball Cocks

The valve that controls the water level in a water closet tank is a ball cock, which is a float-controlled valve (see *Figure 13*). Manufacturers provide repair kits, and you must refer to their specifications and instructions when making repairs. If the stem, valve body, or float mechanism is damaged, you may have to replace the entire float valve. Replacement parts are available for leaking floats or broken float rods.

3.5.0 Tank Flush Valves

Tank flush valves are available in many styles (see *Figure 14*). If the valve and the lever that operates the valve are badly corroded, replace both parts. However, most problems occur with the component parts—the tank ball, the flapper tank ball, the connecting wires or chains, and the guide. Inspect all of these parts to determine whether to repair or replace the entire assembly. If the valve seat is corroded, you can use a reseating tool to restore it.

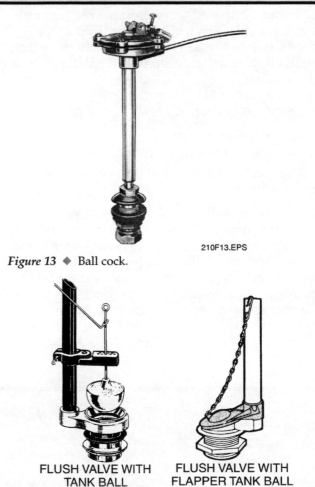

210F13.EPS

Figure 13 ◆ Ball cock.

FLUSH VALVE WITH TANK BALL

FLUSH VALVE WITH FLAPPER TANK BALL

210F14.TIF

Figure 14 ◆ Tank flush valves.

3.6.0 Cartridge Faucets

You can identify a cartridge faucet by the metal or plastic cartridge inside the faucet body. Many single-handle faucets and showers are cartridge designs. Replacing a cartridge is a fairly easy repair that will fix most faucet or shower drips (see *Figure 15*).

3.7.0 Rotating Ball Faucets

A rotating ball faucet has a single handle. You can identify it by the metal or plastic ball inside the faucet body (see *Figure 16*). Drips usually occur as a result of wear on the valve seals. Replacement valve seals and springs are readily available. Leaks from the base of the faucet are usually caused by worn O-rings. You should replace these parts.

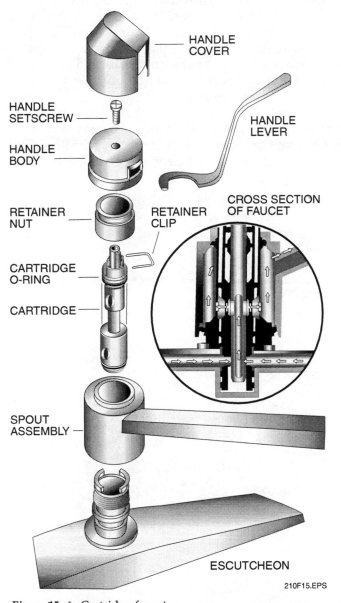

Figure 15 ◆ Cartridge faucet.

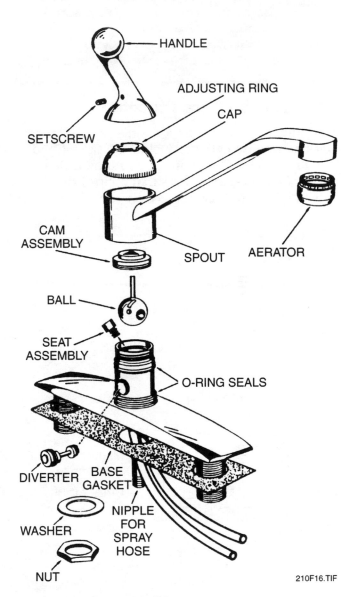

Figure 16 ◆ Rotating ball faucet.

ON THE LEVEL

Replacing Cartridges

Faucet cartridges come in many styles, so you may want to take the old cartridge to the supply house for comparison when you have to replace one.

3.8.0 Ceramic Disc Faucets

A ceramic disc faucet is a single-handle faucet that has a wide cylinder inside the faucet body (see *Figure 17*). The cylinder contains a pair of closely fitting ceramic discs that control the flow of water. The top disc slides over the lower disk.

These discs rarely need replacing. Mineral deposits on the inlet ports are the main cause of drips. Cleaning the inlets and replacing the seals fixes most drips.

4.0.0 ◆ PROBLEMS CAUSED BY INSTALLATION AND WATER

Sometimes leaks occur because of an improper installation. Tub **spuds** that connect the waste and overflow piping and basket strainers on sinks will leak if not tightened properly, if the gasket is not in place, or if you don't use the right amount of plumber's putty. Leaks on waste and overflow tube joints can occur if you don't properly tighten nuts or if you cross-thread them.

Mineral deposits in the water supply can build up in showerheads, aerators, and spray diverters. This clogs the openings and decreases the flow rate. You can take apart and clean most of these fixtures.

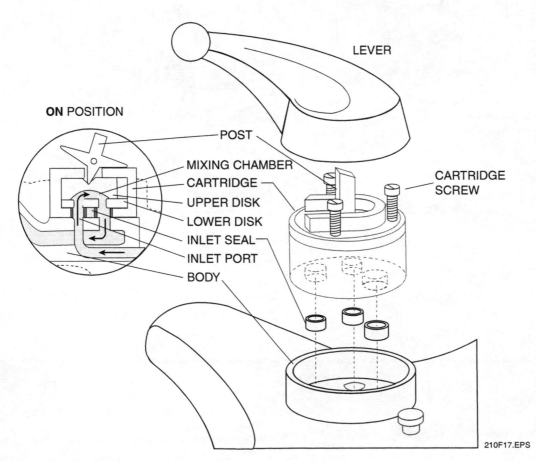

Figure 17 ◆ Ceramic disc faucet.

Summary

Customers may not spend much time thinking about fixtures, valves, and faucets, but they rely on them to operate the plumbing efficiently. When any of these parts stop working you'll be called in to repair, or service, these parts. You will have to minimize damage from major leaks. You must also work safely to prevent injuries to yourself and to co-workers. Finally, you must develop the judgment you'll need to decide when to make repairs and when to replace a fixture, valve, or faucet entirely.

Fixtures include water closets, urinals, and tub and shower drains and valves. Valves come in a wide variety of sizes and styles, but are divided into eight categories for repair work purposes— globe valves, gate valves, flushometers, ball cocks, tank flush valves, cartridge faucets, rotating ball faucets, and ceramic disc faucets. Types of faucets include cartridge faucets, rotating ball faucets, and ceramic disc faucets.

Because valves and faucets operate under pressure and have moving parts, they can wear out or break down. By understanding how valves work, you can repair them and get more use from them, although you'll sometimes have to replace the valve. In some cases, you'll use special tools. In other cases, you'll use a kit supplied by the manufacturer to replace broken parts.

Review Questions

Sections 2.0.0–4.0.0

1. According to the Occupational Safety and Health Administration (OSHA), _____ constitute the majority of general industry accidents.

   a. cuts, lacerations, and bruises
   b. slips, trips, and falls
   c. electric shocks and burns
   d. head and hand injuries

2. Before attempting to repair a leaking valve or faucet, it is recommended that you _____.

   a. place packing around the leak
   b. wrap the leaking fixture with Teflon tape
   c. shut off the valve upstream from the leak
   d. remove any traps near the leaking fixture

3. _____ is a common cause of leaking around the stem or under the knob of a globe valve.

   a. A worn or damaged seat disc
   b. A worn or damaged seat
   c. Worn threads on the stem
   d. Defective packing

4. _____ is necessary to repair a globe or angle valve that rattles when the valve is opened and water is flowing.

   a. Rethreading the stem
   b. Replacing the disc and resurfacing the seat
   c. Repacking and tightening the stem
   d. Replacing the entire valve

5. If the correct size and shape of pre-formed packing is not available, you can use _____.

   a. rope packing
   b. corded packing
   c. twine packing
   d. twist packing

6. If a stem is difficult to turn because of damaged threads you should _____.

   a. replace the valve
   b. repack the stem
   c. install a new packing nut
   d. resurface the seat

7. If a flushometer is leaking at the handle, one possible cause is _____.

   a. a ruptured or damaged diaphragm
   b. high water pressure or volume
   c. worn packing
   d. an improperly seating relief valve

8. If a flushometer is flushing too long, the most likely cause is _____.

   a. the relief valve is not seating
   b. the diaphragm is ruptured
   c. the control stop is malfunctioning
   d. inadequate water pressure

9. _____ may be required to repair a corroded valve seat in a water closet.

   a. A new tank ball
   b. A reseating tool
   c. An entire new assembly
   d. Readjustment of connecting wires or chains

10. Drips in ceramic disc faucets can be repaired by _____.

    a. replacing the seals
    b. replacing the discs
    c. repacking the faucet
    d. installing a globe faucet

Trade Terms Introduced in This Module

Ball cock: A float-controlled valve with a round float.

Cartridge faucet: A faucet that has a metal or plastic cartridge located inside the valve body between the spout assembly and the handle.

Ceramic disc faucet: A single-handle faucet with a cylinder that contains two closely fitting ceramic discs.

Control stop: A valve that regulates the flow of water entering the flushometer. It also allows shutoff of the water supply to individual fixtures.

Flushometer: A flushing device that is connected directly to the water supply and requires no storage tank. It allows repeated, rapid flushing without waiting for a storage tank to refill.

Gate valve: A flow control valve that has a wedge-shaped gate, or disc, that can be raised to allow full flow or lowered to restrict flow. Not intended for very tight shutoff.

Globe valve: A flow control valve with a movable stem that lowers to a fixed seat to restrict flow. The stem has a washer to provide tighter closure. The valve body is shaped like a globe.

Packing extractor: A tool used to remove packing from a valve stem.

Packing nut: A nut that holds the packing around a stem in place.

Removable seat wrench: A tool used by plumbers to replace valve and faucet seats.

Reseating tool: A tool used by plumbers to reface a valve seat.

Rotating ball faucet: A single-handle faucet that has a metal or plastic ball inside the valve body.

Screw extractor: A tool designed to remove broken screws.

Spud: A short pipe that serves as a connection in a piping system.

Tank flush valve: A valve that regulates the flow of water in a water closet tank.

Tap: A tool used for cutting internal threads.

Twist packing: A string-like material used to pack valve stems when pre-formed packing material is not available. It is not as durable as pre-formed packing.

Valve bonnet: A valve cover that guides and encloses the valve stem.

Wire drawing: A defect that occurs in gate valves when they are opened only part way. The disc tends to vibrate, which causes the edge to erode.

Answers to Review Questions

Answer	Section
1. b	3.1.0
2. c	2.1.0
3. d	3.1.0
4. b	3.1.0
5. d	3.1.0
6. a	3.1.0
7. c	3.3.0
8. a	3.3.0
9. b	3.5.0
10. a	3.8.0

ACKNOWLEDGMENTS

Figure Credits

Coyne and Delany Company	210F01, 210F02
Crane Company	210F14
Crest/Good Manufacturing Company	210F04, 210F05, 210F08, 210F11, 210F13
Mansfield Plumbing Products	210F12
NIBCO	210F03
J. A. Sexauer Manufacturing Company	210F06, 210F07, 210F09

NCCER CRAFT TRAINING USER UPDATES

The NCCER makes every effort to keep these textbooks up-to-date and free of technical errors. We appreciate your help in this process. If you have an idea for improving this textbook, or if you find an error, a typographical mistake, or an inaccuracy in the NCCER's Craft Training textbooks, please write us, using this form or a photocopy. Be sure to include the exact module number, page number, a detailed description, and the correction, if applicable. Your input will be brought to the attention of the Technical Review Committee. Thank you for your assistance.

Instructors – If you found that additional materials were necessary in order to teach this module effectively, please let us know so that we may include them in the Equipment and Materials list in the Instructor's Guide.

Write: Curriculum Revision and Development Department
National Center for Construction Education and Research
P.O. Box 141104, Gainesville, FL 32614-1104

Fax: 352-334-0932

E-mail: curriculum@nccer.org

Craft _____ Module Name _____

Copyright Date _____ Module Number _____ Page Number(s) _____

Description _____

_____

_____

_____

(Optional) Correction _____

_____

_____

(Optional) Your Name and Address _____

_____

_____

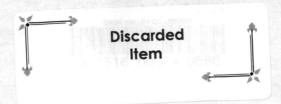